RIVER FRIEND

A series of Riverine Small Books

by Sylvia M. Haslam and Tina Bone

BOOK 9

STREAM STORY II: Another Riveting Riverscape
River Cam, Cambridge

(RIVER CAM [Cambridge] and LOWER GREAT OUSE)

Baits Bite Lock Weir, The River Cam, Horningsea

Ninth book published in the Series:

STREAM STORY II: Another Riveting Riverscape—River Cam, Cambridge
[RIVER CAM, Cambridge and LOWER GREAT OUSE]

A Book in a series of Riverine publications by

Sylvia M. Haslam and Tina Bone

(Each book is about a different subject so the series can be read in any order)

Written and Edited by Sylvia Haslam and Tina Bone.
Illustrated by Tina Bone (unless otherwise stated)

Copyright © 2025:
S. M. Haslam and T. Bone

All rights reserved. No part of this book may be reproduced, stored in a retrieval system, or transmitted in any form or by any means, electronic, electrostatic, magnetic tape, mechanical, photocopying, recording or otherwise, without permission in writing from the Authors (email: ourbooks@tinasfineart.uk).

RFS9: PAPERBACK **88** pp.
ISBN No. 978 1 9162096 9 5
82 Illustrations

Published by: Tina Bone UK
First edition: **December 2025**

https://riverfriend.tinasfineart.uk

Email: ourbooks@tinasfineart.uk

Contents

Introduction 1
Earthworks 11
The Fenland 16
The Main Streams of the Upper River Cam 19
   Bourn Brook 20
   River Rhee 22
   Upper River Cam or River Granta—usually known now as the River Cam . . .29
Upstream of Cambridge 32
Cambridge 37
Place Names 48
River Settlements 49
Uses of rivers 53
   Water 53
   Fishing 53
   Milling 54
   Fish Ponds 54
   Transport 55
   Recreation 55
   Legal Navigation 56
   Water Fowl 56
   Thatch 56
   Local Enterprise 56
River towns 57
Lodes 61
   Lodes villages 61
   Lodes (Roman canals) 62
Vanishing Land 66
Two pollution stories 68
   1. Upstream of Cambridge at Harston 68
   2. The River Cam in Central Cambridge 68
Downstream of Cambridge 69
   Ely 71
   Littleport 74
   King's Lynn 75
Citations 82
List of Published Stand-alone Titles in the River Friend Series 83
About the Authors 83

INTRODUCTION TO THE SERIES

Rivers are vital. They bring freshwater to the land, on which all its life depends. They are beautiful and fascinating, making up both the typical British countryside and many of its most spectacular views. If they vanished, what hardship and outrage there would be! Yet, slowly, slowly, they are vanishing, the larger stream becomes smaller, the tiny brook becomes a ditch and dries, and is filled in—the small ditches get polluted and dug out, become dull, and vanish from sight and consciousness. How can we save our rivers and riverscapes? How can we raise awareness on this slow, almost invisible loss?

We believe that this series of handy, small books, suitable for readers from teenage upwards, will help to raise awareness. Individually, each book tells a story on a particular riverine and riparian environment. Collectively, the series will inform the extraordinary value of freshwater and its plants, as well as natural heritage.

The Authors realised that there was a huge gap in the literature. There are many publications for scientists, for pond-dippers, birders and anglers, but handy pocket books focussing on the river itself, and the vegetation belonging to it and creating the habitat for all else: we could find none!

For explanations regarding British freshwater plants, terminology mentioned throughout the series, and Picture Guide and reference section for further reading, see the book entitled *A PROLOGUE TO THE SERIES: Plant identification and Glossary of Terms* (also available to view in pdf format free on-line at https://riverfriend.tinasfineart.uk/resources/).

Other titles in the Series are listed on the last page of this book and on the River Friend Website:
https://riverfriend.tinasfineart.uk

STREAM STORY II: Another Riveting Riverscape
River Cam, Cambridge

(RIVER CAM [Cambridge] and LOWER GREAT OUSE)

Introduction

Is this booklet about another riveting riverscape? Yes indeed. *The River Brue, Somerset* (Stream Story II, Book 2 in the River Friend Series) is truly riveting as a western, west-flowing river, with hills, lowlands, wetlands, a complex history, wonderful architecture, cathedral with large and prosperous mediaeval lands, and magic. Remove the hills and magic, add **Denver Sluice** and the **Roman Lodes** and the River Cam is just as riveting! Both rivers were major waterways for boat transport, the downstream ports having major navigation with significant trade inland to British ports and to the continent. And, of course, both had many water mills—most initially for milling grain into flour but later diversifying greatly, especially after the workforce was considerably diminished by the Black Death which killed up to half the population in the 1340s.

The River Cam drains a surprisingly large area of the East Midlands all of which (upstream of Cambridge) is food-producing, agricultural lowlands (Fig.1), There are no mountains, the highest "hill" being the "chalk downs" in the Royston, River Rhee area at 168m (551 feet) above sea level. There is remarkable little industry in the catchment. The most southwards stream (River Cam proper) rises only a few miles north of Stansted Airport. There are a surprising number of rivers with sources within about 8 miles of this airport. "Surprising" because the land is neither mountain, nor wetland, nor an area of many springs. It has, though, tributary sources of the River Chelmer, River Stort, and River Quin, apart from that of the River Cam.

Downstream of Cambridge the River Cam shortly flows into The Fenland where it stays for most of its journey to the sea at King's Lynn, after joining the River Great Ouse at Pope's Corner a few miles from Ely (Fig. 2). The Fenland is a former wetland, the soil being mostly fen peat. Fen peat is dead, plants, so it is organic-rich (with invertebrates and microorganisms where possible). On land, vegetation, if not eaten, is mostly broken down or oxidised with a little remaining in the soil. Fen peat is dark, near-black when exposed to air, and this area is also known as The Black Fen. Fen peat grows under water. Because the water is mostly run-off from the agricultural lowland it is fairly high or medium in nutrients, with some chalk upstream influence, so is the peat. Fen peat thus differs from bog peat which is fed by nutrient/calcium-poor rain and is built up on land, not under water. In The Fenland there are some islands of clay, the Isle of Ely being the largest. There are some other variations. In the west, there is some more nutrient-low, semi-bog peat. In the east, near the sea, there is more inorganic silt and mud, with more again in the north Fenlands.

The River Cam, as here described, is long, rising at its furthest source near London (in Essex) and extending down through agricultural lowlands, just touching the small area of chalk downs in Hertfordshire (near Baldock). It then reaches the ancient port, market, and town (later University city) of

Fig. 1a. The River Cam showing its tributaries, villages and towns (Resource courtesy of the East Anglia Area Environment Agency)

Cambridge. Here there was even foreign shipping. Downstream, it is joined by the Great Ouse before reaching upstream of Ely, and so passes to the North Sea, now via King's Lynn.

The coming of the railways in the nineteenth century gradually took the trade away from the river, rail freight being much easier inland than boat cargo. In its heyday, the Great Ouse drained much of the East Midlands, from the Cam to near Barnwell and Rushden. All larger rivers carried ships as well as small boats and later barges; the smallest rivers carried small boats. The River Cam formed a very important means of communication all over the eastern parts of England, more so for example than in the Somerset River Brue area.

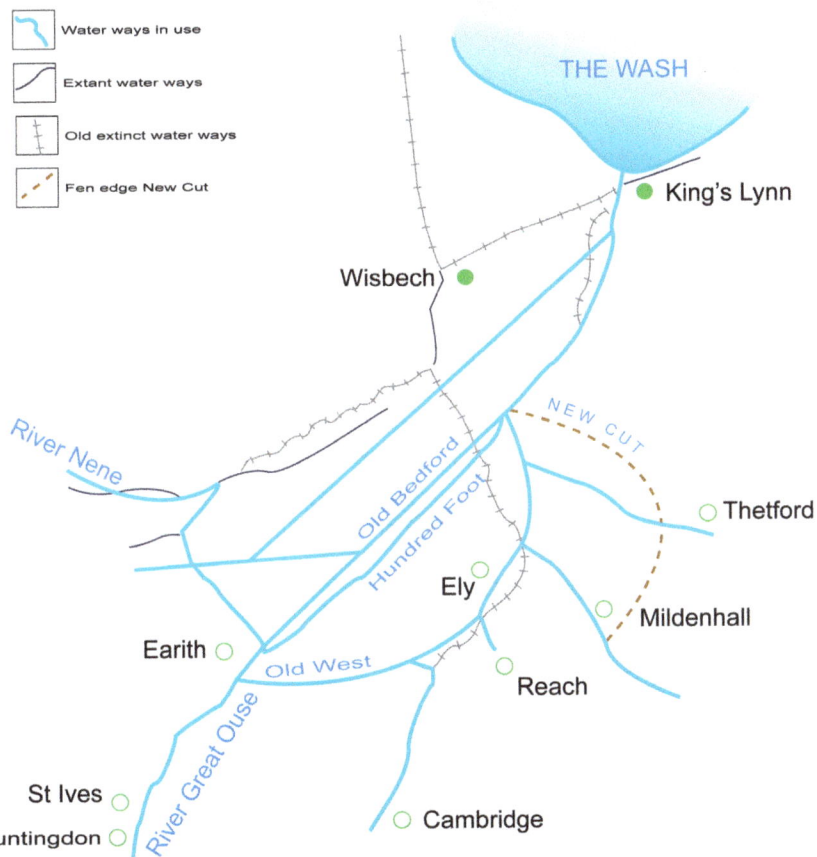

Fig. 1b. Schematic Great Ouse historical river plan (re-drawn from Haslam, The Historic River). Hatched lines old extinct waterways and coast line, blue lines, present waterways, drained and straightened since the move of the outfall from Wisbech to Kings Lynn, The dashed line represents the recent fen edge New Cut

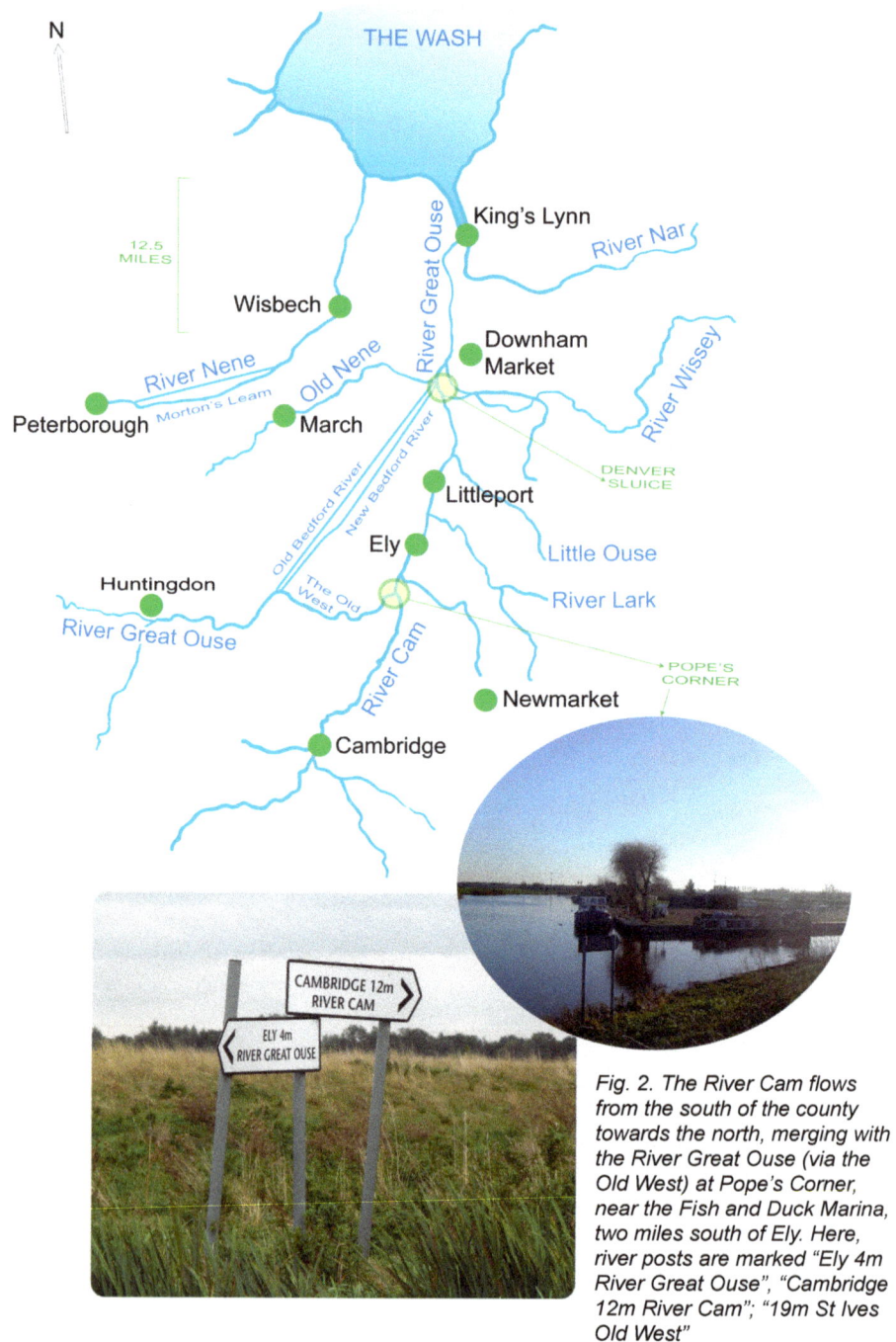

Fig. 2. The River Cam flows from the south of the county towards the north, merging with the River Great Ouse (via the Old West) at Pope's Corner, near the Fish and Duck Marina, two miles south of Ely. Here, river posts are marked "Ely 4m River Great Ouse", "Cambridge 12m River Cam"; "19m St Ives Old West"

Even the more limited River Cam drainage basin has far too much of interest to describe all the places and parts worthy of the reader's attention and, like the book about the River Brue, this book selects some and omits most. That does not mean it selects only the most important—however "important" should be defined. There are other papers and pamphlets—ones from most of the "watery" organisations, for example, the Environment Agency, World Wildlife Fund, Rivers Trust, Cam Valley Forum, Wildlife Trusts, Wetlands and Wildlife Trust, Chilterns Area of Outstanding Natural Beauty, Salmon and Trout organisations and the Wild Trout Trust. These agencies are well worth consulting, and anyone who finds this booklet interesting should read all the others!

Coming into the fens just downstream of Cambridge is the village of Stow-cum-Quy. In the mid-twentieth century black soil—fen peat—started just below the village because when the peat was being laid down there was flood here for most or all of the year. Now, drive down the gentle slope and the River Cam is almost in sight before the fields are entirely black. Loss of peat means the less fertile alluvium underneath is now the "soil". So there is a loss of food production and a loss of pollen which by analysis shows what vegetation was there at what time in the past.

Sea and land levels have varied a lot over time. Periods with really good food production (dry land, warmer) were Roman, mediaeval, and post-seventeenth century (warmer only recently). The first two dry periods were "natural". The recent one is man-made following the draining of the fens, the major mechanical construction of which was Denver Sluice with its later improvements and replacements.

Denver Sluice is now a massive and complex construction (Fig 3)! It drains The Fenland and prevents sea water from The Wash backing up into the main rivers. There is also the occasional excitement of a tidal bore which usually occurs when there is a large rise between high and low tide (Fig. 4).

Because so much peat has vanished, The Fenland is now lower than where rivers entered it, as shown in Fig. 5. Deplorable, but fascinating! Denver Sluice maintains the agriculture on the low Fenland, the main ground level. Now, the well-publicised "Climate Change" of the past two centuries is coupled with water loss (drainage and abstraction) to cause major alterations.

People living in the lowlands supported themselves mostly by farming. Available crops were few, the importance of fruit and vegetables not properly recognised, plant breeding not efficient, and fertiliser was sparse. Even after the eighteenth-century Enclosure Acts, the same limitations applied, and in the absence of potatoes (seventeenth century), bread was even more important: made from wheat, but also from oats and barley.

Village crops were badly affected by the Enclosure Acts (the proper compensation was decreed by those Enclosing) but before the Post Office Savings Bank existed, there was nowhere to invest the capital value of, say,

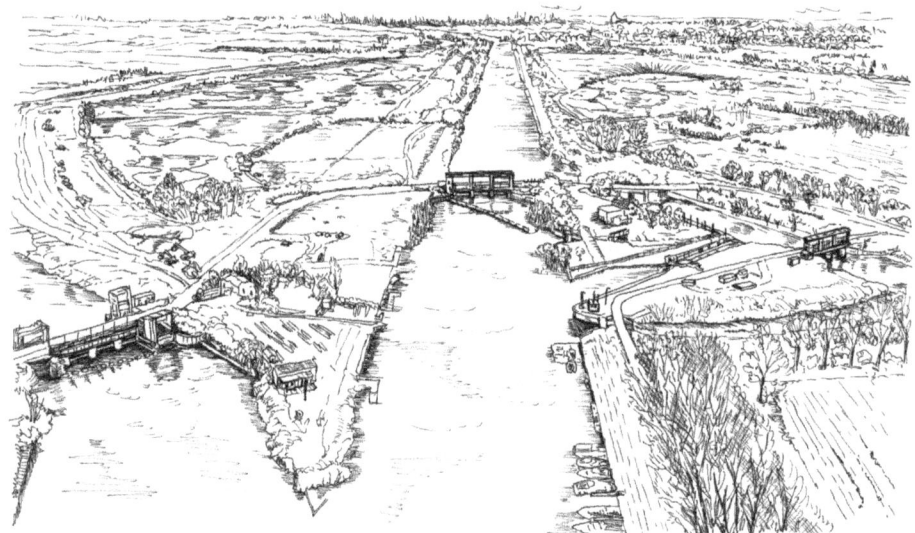

Fig. 3a. The Denver Sluice Complex sits at the confluence of several watercourses which include the River Great Ouse, New Bedford River, or Hundred Foot Drain, River Old Nene and River Wissey

Fig. 3b (below). Closer view of one of the sturdy "gates" comprising the Denver Sluice Complex spanning the River Ouse

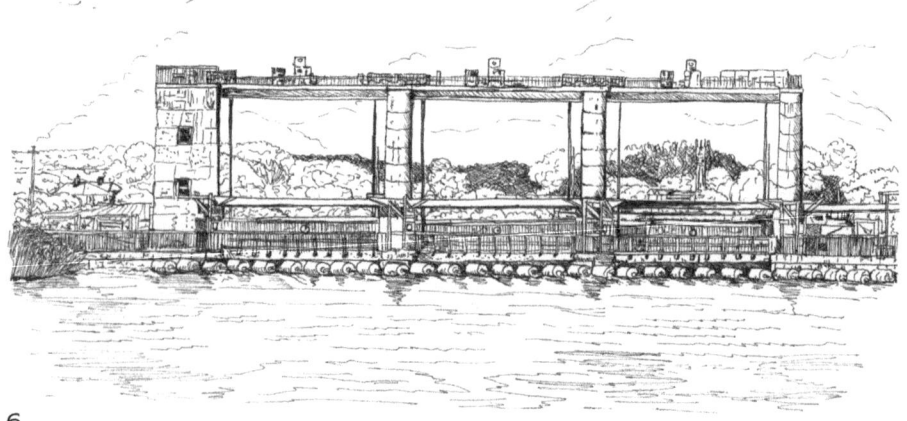

one cow, two geese and ten chickens feeding on the Open Field. Spent as income, this led to malnutrition in years of bad harvest. The depressing hymn written for such years included:

> *What our Father does is well,* | *Though He sadden hill and dell* | *...Though nor milk nor honey flow,* | *In our barren Canaan now,* | *God can save us in our need.*

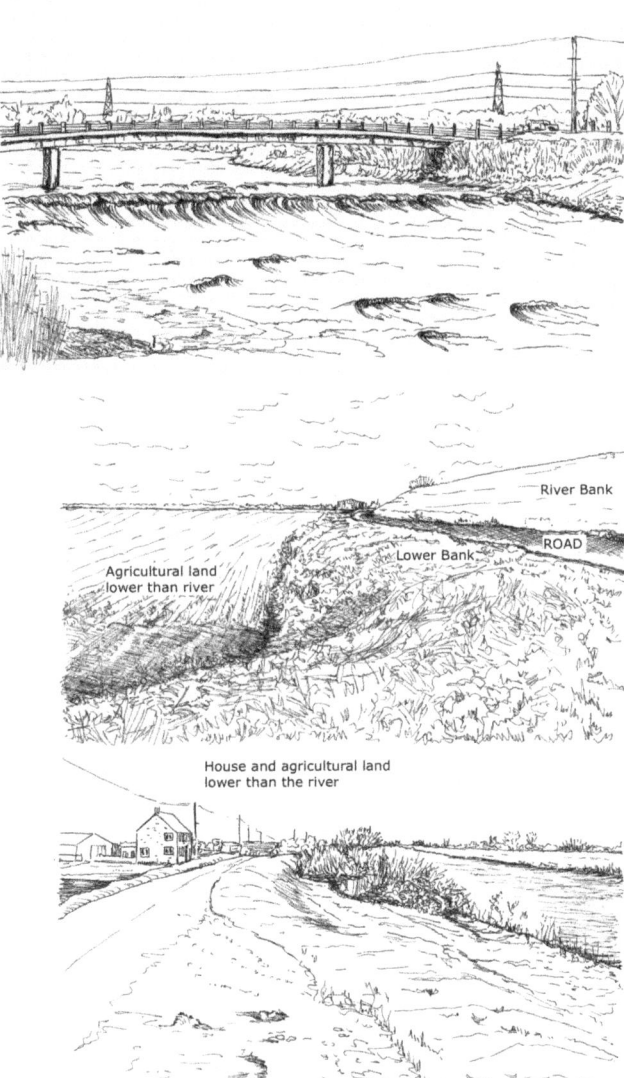

Fig. 4. A seasonal tidal bore running up the River Great Ouse from The Wash in April 2021. Resource ITV Anglia News

Fig. 5. Two aspects along the River Great Ouse embankment (Ten Mile bank) showing the road(s), agricultural land, and dwellings below the level of the river, nominally 1.8m (6ft) below sea level

Enclosed Fields were better managed so produced more food, but the relief of these local famines came more from better transport: first the eighteenth-century canals and toll roads; then the nineteenth-century railways. Food could be sent cheaply throughout the country and this included transport around the River Cam.

Grinding corn by quern was time-consuming and women doing this had little time for dairy, hens, ducks, fruit, still-room (medicines, drinks, etc.), dressmaking, as well as family and house (Fig. 6). The answer was mills. Water mills were brought in by the Romans in the first century BC. They quickly spread throughout the country and were listed in *Domesday Book* — the extraordinarily detailed description of England ordered by William I, showing his recently-conquered realm (1086).

Fig. 6. Grinding corn with a quern dates back to Neolithic times. Typically comprising two stones: the bedstone and upper handstone, the upper stone is moved by hand in a circular motion to crush and grind the grain

Mills could be turned "undershot" by the main flow of a lowland stream, or "overshot" if the leat was raised so that the flow fell onto the wheel. Overshot mills were mostly used in the hills utilising a stream's fast, natural fall and flow, although in flat lands an overshot wheel was possible needing dams and an elevated head race—much more costly than using the less forceful, undershot natural-level stream flow (Fig 7).

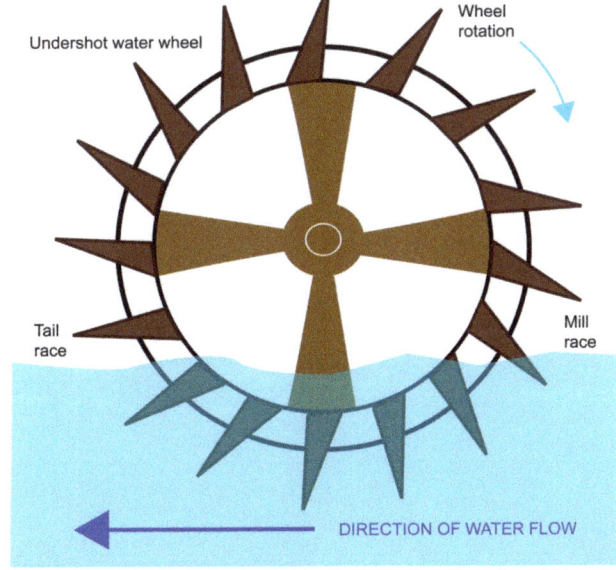

Fig. 7. The simplicity of overshot and undershot mill wheels. The Overshot system is more efficient as more energy is harnessed as the water drops into buckets; undershot wheels rely on the natural speed of the water flowing underneath

Many complaints are recorded of the Landlord charging extortionately for the use of the village mill, and requiring tenants to use it. It is rarely said that a mill is expensive to build and run, and if the Landlord has built one, it is reasonable that tenants take their grain there. The Landlord himself could usually pay transport for his grain to the next village mill; could the villagers? So a mill was a useful asset to a village.

Along the Upper Cam, most villages had at least one mill, for example, Grantchester (close to Cambridge), Great and Little Shelford, Barrington, Whittlesford, and more: in fact wherever there was enough water to turn a wheel. Part of the reason for the Upper Cam Drying Up in recent times has been masked by an unsung, man-made change—namely the loss of mill-gates, weirs, and similar obstructions to water flow. Cross-river obstructions like those in Figure 8 hold up water above, and let it run shallow and swifter below. Often mills were so frequent that the deep, slow water from one weir stretched back to the next one upstream (top-to-toe). As steam, then electricity or oil, took over the power which had previously been supplied by water mills, the "obstructions" were often just left in the river. By the 1980s and beyond,

Fig. 8. View looking from the white footbridge at Baits Bite Lock Weir, The River Cam, Horningsea. Lock gait for boat passage is to the right but not shown

however, "tidying up" removed most of them and they were replaced with low weirs which did not deepen the river. The effect on the vegetation was deplorable and also contributed to yet more Drying Up.

From the 1970s, disintegrating or unwanted river structures were gradually removed. New ones were likely to be low weirs with stakes to measure water level or, later, permanent electric gauges. Either way, dropping water levels and lessening water is usually bad for the river. Big fish need big waters, no water means no aquatic life. Added to the Drying Up from other causes such as abstraction and drainage is this loss of water by shallowing the rivers. Mills were mostly in place by $c.$ 1500AD, so where these were plentiful it is more than dubious to say "natural rivers" do not have mills! (What habitats have good modern records going back to 1500?)

Fishing, for all the edible fish, took place at any suitable spot, and is described below.

Earthworks

Earthworks (Anglo-Saxon or earlier) were anciently called "Dykes". These separated countries or parts thereof. The best known and Britain's longest ancient monument is Offa's Dyke in the Welsh Borders. (Dykes were embankments, dug out so that the great ditch faced the expected enemy who, having to descend, was both hampered as well as vulnerable to attack by soldiers on the bank above. Three such dykes run from the south to the south edge of The Fenland where they link to lodes: Fleam Dyke, from Balsham to Fulbourn (Fig. 9a); Devils Ditch or Dyke. Its height can be up to 10m and it runs from Wood Ditton to Reach (where it ends near the "Dyke's End" pub), then on to join the River Cam (Fig. 9b); and Car Dyke (Fig. 9c), which runs from Cambridge to Landbeach to the Great Ouse and River Nene, and on up into Lincolnshire.

Nomenclature of channels is difficult. Early earthworks are DYKES, the sunken channels having this military purpose, but they did not necessarily contain water. More recently, innumerable straight, man-made channels criss-crossed The Fenland, the larger or more important, when dug out, had the debris put on one bank, making a track above the wetland level—a useful road. The two together (road plus water channel) were DYKES. The "roads" were so awful that it was said that, if the idea of hell had developed in

England, it would not be a burning fiery furnace but to be lost in The Fenland in fog (common in genuine wetland), frost, and a strong east wind. In the twentieth century, with roads made properly but sparsely, an English dyke gradually became just a channel, and the name started to take over in other wetlands to describe channels larger than a DITCH ($c.$ 0.5–2m wide) and smaller than a DRAIN ($c.$ 8–12+m wide). Meanwhile, in Scotland DIKE came to mean drystone walls in the countryside. In The Netherlands, DIJK, meaning embankments (often sizeable), usually included the channel alongside, usually with water. (In the US, the variable naming presumably depended on where in Europe the early settler farmers came from.)

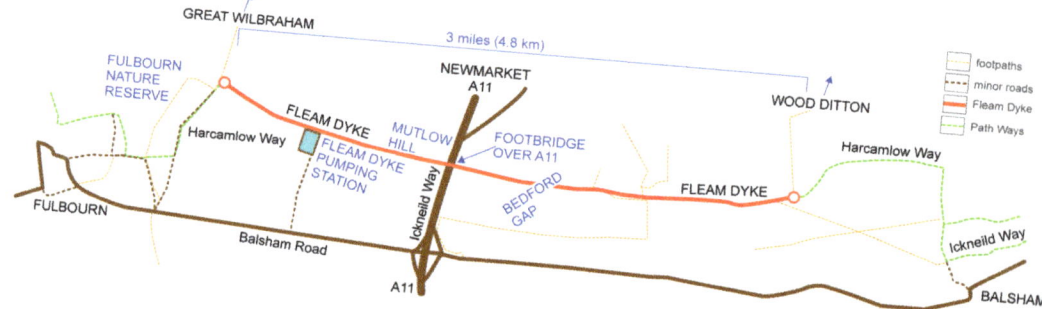

Fig. 9a. Fleam Dyke is 3 miles in length (4.8km). The Schematic route map shows the full length of the dyke, which passes over the A11, from Fulbourn to Balsham, with several noted landmarks

9b(ii)

9b(i)

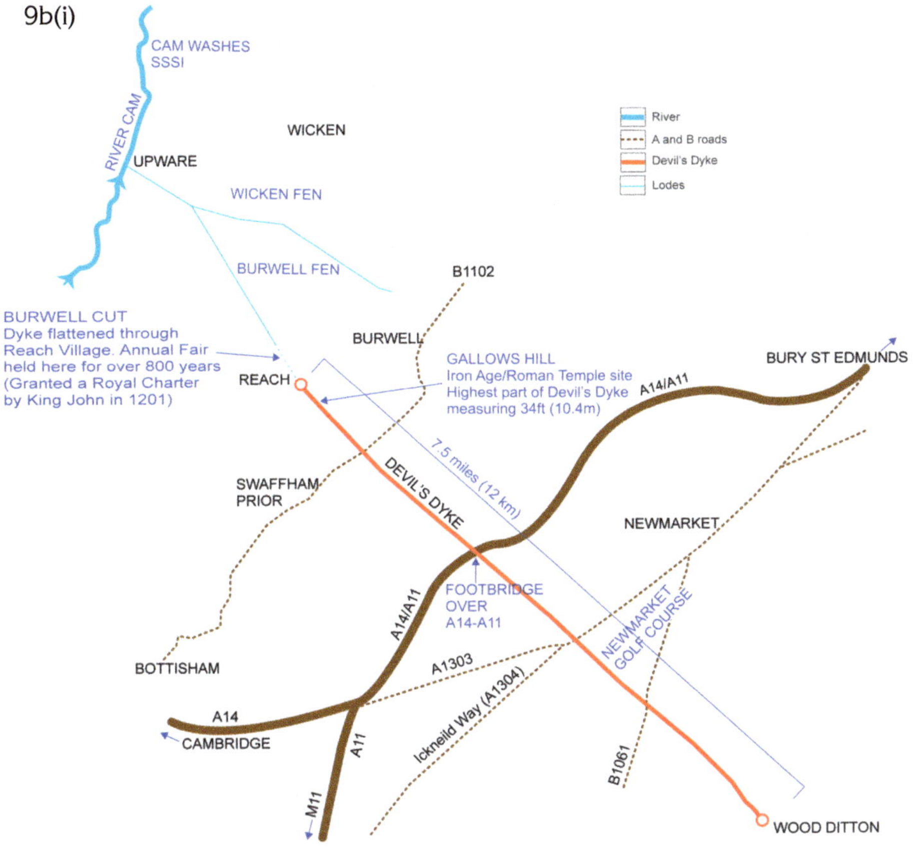

Map labels:
- CAM WASHES SSSI
- RIVER CAM
- UPWARE
- WICKEN
- WICKEN FEN
- BURWELL FEN
- BURWELL CUT
- B1102
- BURWELL
- GALLOWS HILL — Iron Age/Roman Temple site. Highest part of Devil's Dyke measuring 34ft (10.4m)
- REACH — Dyke flattened through Reach Village. Annual Fair held here for over 800 years (Granted a Royal Charter by King John in 1201)
- A14/A11
- BURY ST EDMUNDS
- SWAFFHAM PRIOR
- DEVIL'S DYKE — 7.5 miles (12 km)
- NEWMARKET
- FOOTBRIDGE OVER A14-A11
- A14/A11
- NEWMARKET GOLF COURSE
- BOTTISHAM
- A1303
- Ickneild Way (A1304)
- A14
- CAMBRIDGE
- A11
- M11
- B1061
- WOOD DITTON

Legend:
- River
- A and B roads
- Devil's Dyke
- Lodes

9b(iii)

Fig. 9b(i) (above). Devils Dyke (7.5 miles (12 km): Schematic route of the dyke from Wood Ditton to Reach

(ii) (left) Typical outlook, with pine trees, along the Dyke near Newmarket Golf Course

(iii) (right) Typical dyke landscape near the fen edge, looking towards Gallows Hill

Car Dyke is 57 miles long (92 km). The name "Car Dyke" is thought to derive from the fourteenth century word "Carr", meaning "marsh" or "drained land" (Figs 9c(i), (ii), (iii)). Car Dyke is now considered to be a great engineering feat, alongside Hadrian's Wall which is situated on the border between Scotland and England, and was probably built during the second century AD.

9c(i) This figure shows waterways in the early fourth century which could have been used to link the sections of Car Dyke. Roman Roads of the time are also shown to paint a broader picture linking the lower Cambridgeshire works to the Peterborough–Lincoln Dyke

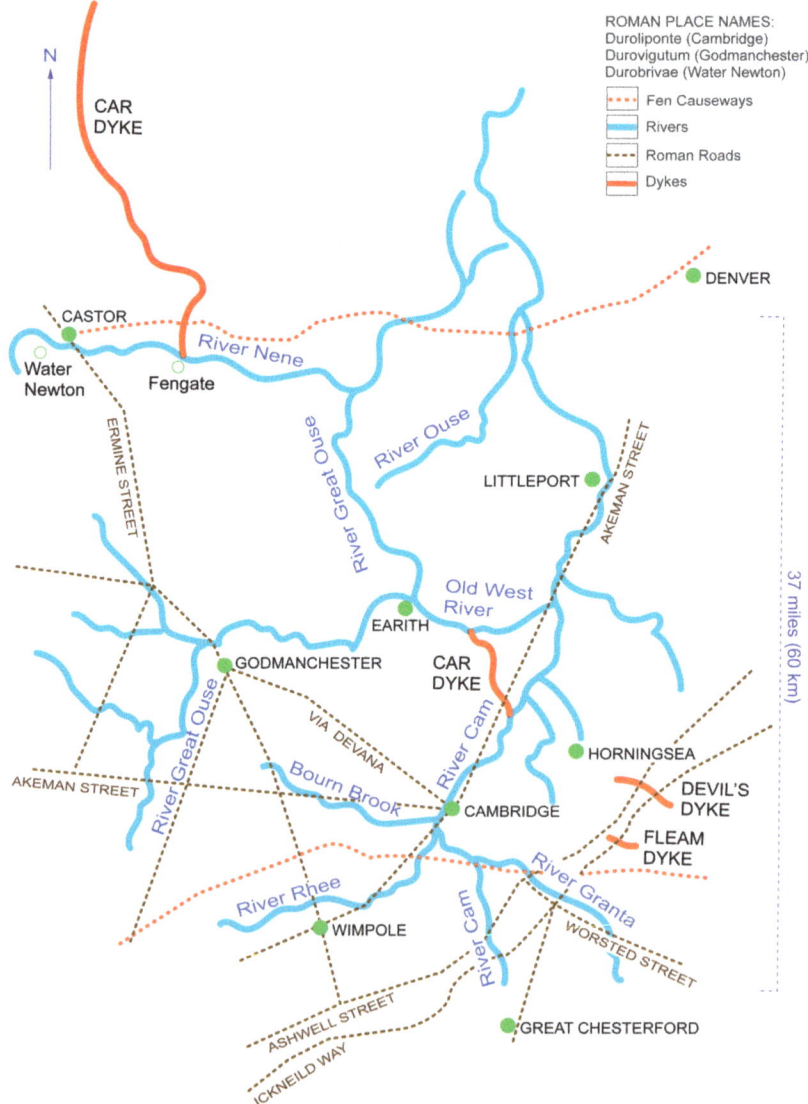

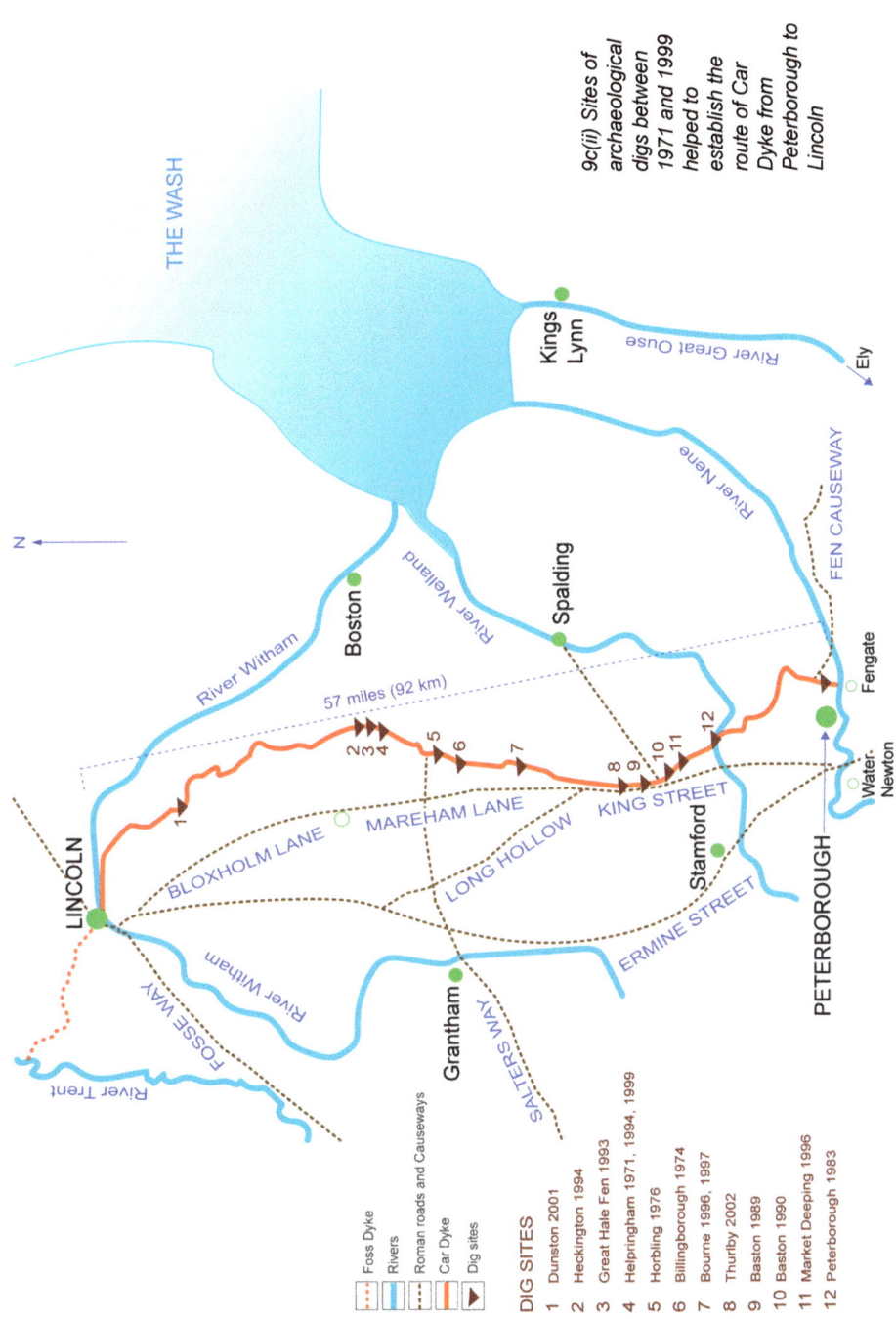

9c(ii) Sites of archaeological digs between 1971 and 1999 helped to establish the route of Car Dyke from Peterborough to Lincoln

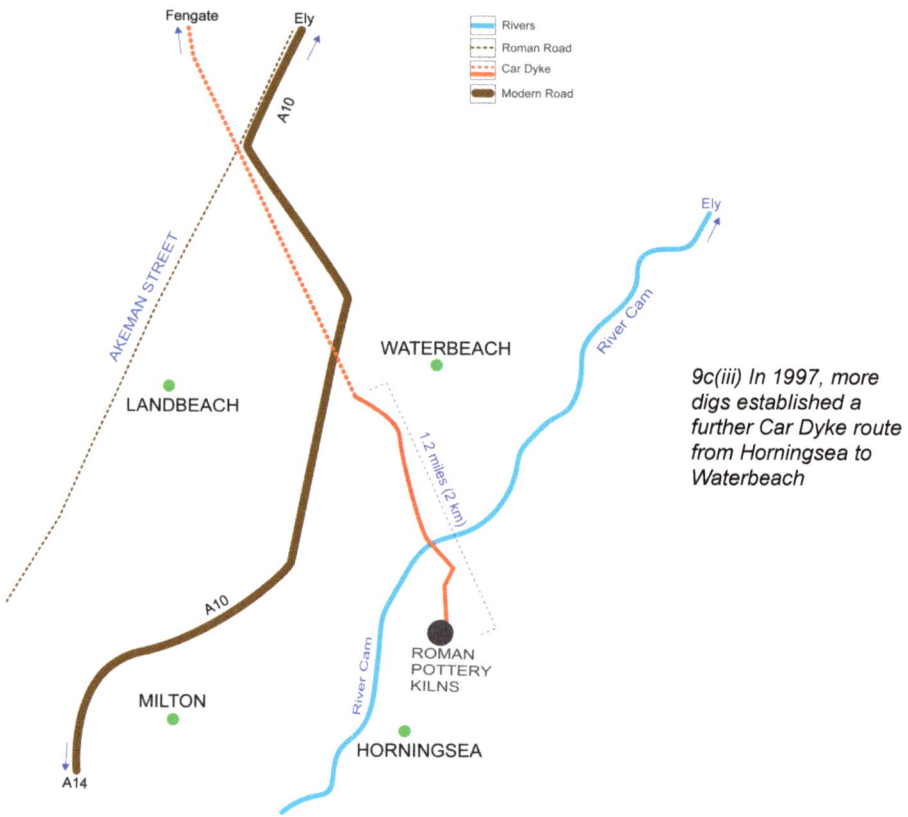

9c(iii) In 1997, more digs established a further Car Dyke route from Horningsea to Waterbeach

The Fenland

The Fenland villages formed a community of their own, heavily dependent on the water and wetland. Crops were limited but fish, including eels, were plentiful. Craft industries were presumably more restricted than in the lowlands, but thatch, walls, furniture, fencing, fish traps, animal bedding, baskets, and such-like could all be made from reeds and sedges, at least in a rudimentary fashion, as shown in the few examples opposite in Figure 10.

FIGURE 10 (OPPOSITE). Typical thatched cottage, pens, baskets and arrows made from reeds. (Learn all about the fascinating and versatile Reed (Phragmites) in Book 5 in this series: REED— On the Edge)

One special craft, however, did flourish: making the fen-runner skate, with an unusually long blade, in contrast with Dutch ice skates of the time (Fig. 11a,b). (Temperatures dropped post-mediaeval, and there was ice for much of the winter.)

Pole vaulting to cross dykes with water presumably flourished also, as in The Netherlands. In 1981 the inaugural British National Dyke-Jumping Championships were held at Ferry Meadows Country Park, Peterborough. Competitors had to pole vault over a dyke which measured 27ft across (8.2m) and the champion was Chris Douse. In recording the event, the BBC mentioned that 14 of the 100 competitors were women, but sadly none made it to the other side, and that before the fens were drained leaping across dykes was a necessity, but was now a sport (Fig. 11c).

Boats were an important part of life, even more so here than in the Low Countries.

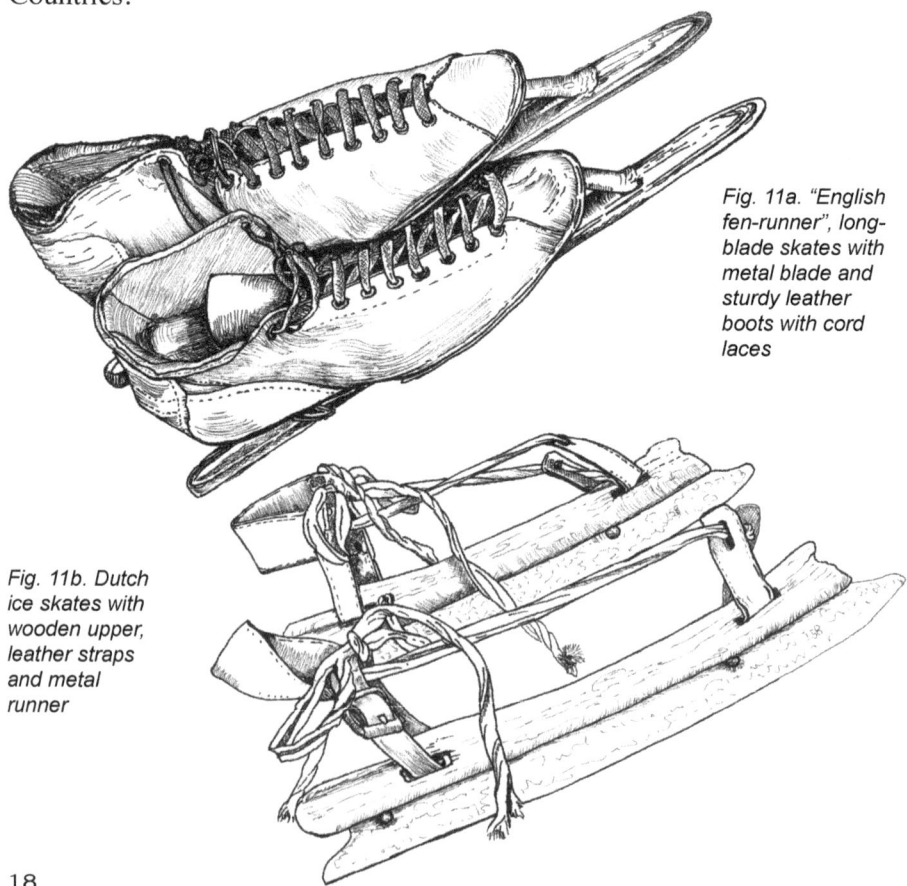

Fig. 11a. "English fen-runner", long-blade skates with metal blade and sturdy leather boots with cord laces

Fig. 11b. Dutch ice skates with wooden upper, leather straps and metal runner

Fig. 11c. Crossing a Dyke by pole vault

The Main Streams of the Upper River Cam (Fig. 12)

Figure 12 shows the River Cam and its larger tributaries (west to east flowing northwards towards The Wash) and the lower part of the River Great Ouse which it joins for the final flow to the sea.

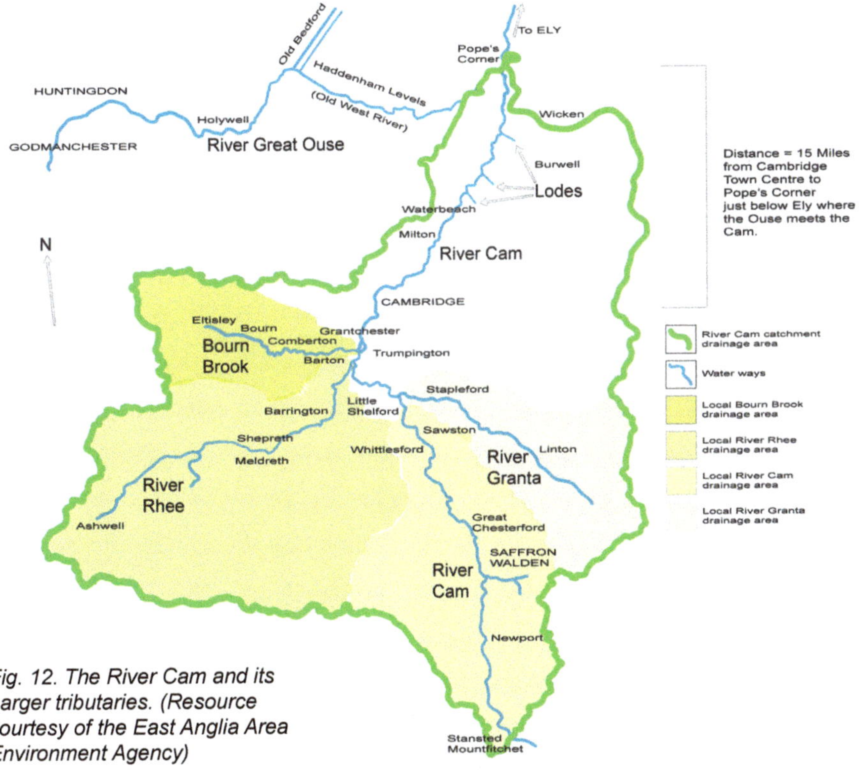

Fig. 12. The River Cam and its Larger tributaries. (Resource courtesy of the East Anglia Area Environment Agency)

Cambridge conveniently divides the Upper Cam—beyond the major navigation and the Lower Cam—it is the major town and the water highway to the sea. Four tributary streams comprise the Upper Cam. Their names are, even after two centuries of Ordnance (ammunition) Survey maps, still a bit fluid. In this book these are referred to as the Bourn Brook, River Rhee, River Cam and River Granta.

Bourn Brook (Fig. 13)

The most northerly and westerly stream is the Bourn Brook—a typical clay stream—dug out (later dredged) and much disturbed along its course. It rises in Eltisley and travels east under the M11 motorway where it joins the River Cam near Byron's Pool (Grantchester). The Brook's aquatic plant species recorded previously when there was more water and less upset, are characteristic of clay, and make a good contrast with the (more chalk) aquatic vegetation of the other three upper tributaries. The contrast between the Bourn Brook (clay) and the River Rhee (chalk) is great, and a standard.

Bourn Brook January 2020, Caxton End ford. Brook is flowing centre of photograph from right to left. Top water is run-off via the road in to the brook from a large agricultural drainage channel further up

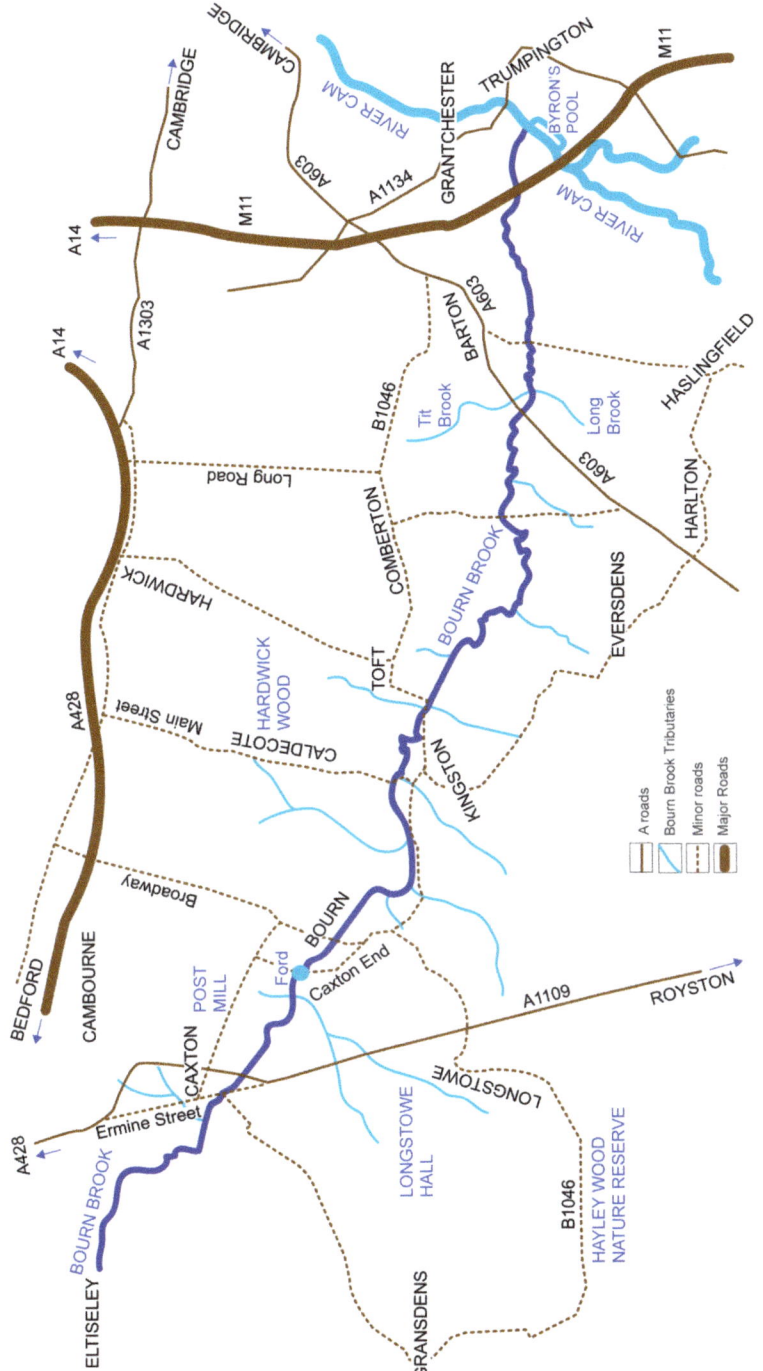

Fig. 13. Bourn Brook, which is 16 miles (25.8 km) long, showing main tributaries from its rising near Eltisley to joining the River Cam at Byron's Pool, Grantchester

21

River Rhee (Fig. 14)

The next river along the arc to the south is the River Rhee, still marked as "or Cam" on Ordnance Survey maps. This is a contrast in the main part—as much of a chalk stream as the Bourn Brook is a clay one. It has a variety of tributaries, including the River Mel (Melbourn, Meldreth) which has the immense advantage since the early 2000s of having an active conservation group willing to get wet and muddy! There are many more such groups in 2025. Commonly the chalk forms Downs (hills) and here, near Royston, are the highest hills of the Cam Valley catchment. Despite all its use (*and misuse*) the River Rhee still flows well and has good chalk *Ranunculus*-based vegetation where disturbance or pollution does not hinder its growth.

Fig. 14 (right). The route of the River Rhee, rising from Ashwell, flows eastwards for 12 miles (19 km), and joins the Cam near Byron's Pool, Grantchester. The map also shows tributaries which flow into the Rhee from the chalk hills near Royston down to Barrington

Fig. 14 (above). Looking downstream, River Rhee at Harston Bridge September 2011. (Good flow, little silt, good Ranunculus)

Fig. 14 (left). Same place as top photograph, but taken in September 2012—note the difference in aquatic bed and bank plant species. (Less flow, less Ranunculus and Apium nodiflorum, more Sparganium emersum)

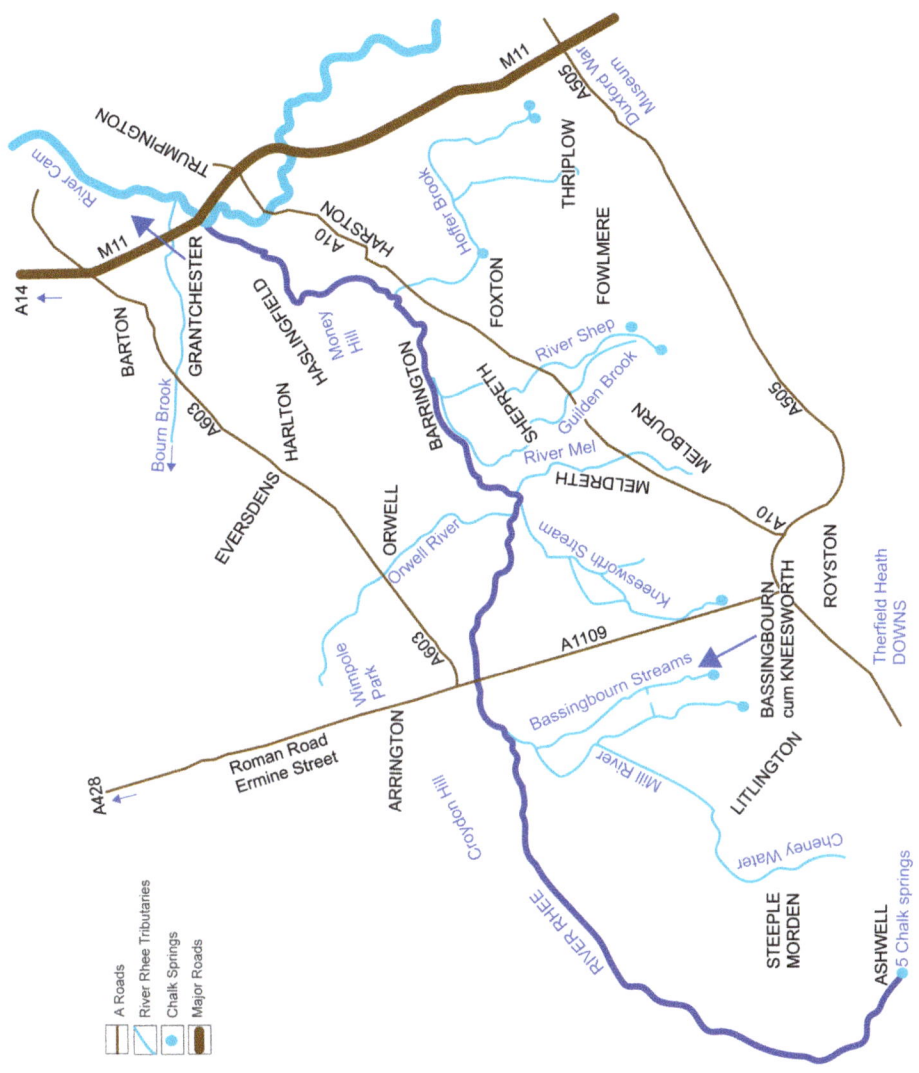

The River Rhee rises on chalk at Ashwell springs near Royston and has (or had) many springs with good outflow, and joins the River Cam, just south of Byron's Pool. Bourn Brook, on the other hand, being on clay, has headwater breaks with run-off and up-seeping water—yet another contrast.

The River Rhee in a now quiet part of England, was not always quiet! Way back, in the ninth century, Danes arrived to conquer and settle. These Vikings came by ship, and not just to the coast, as we would expect, but up the River Cam—and then up the small River Rhee. Remarkable! The River Rhee would still take small boats for transport, but the idea of a main Danish invasion of longships is no longer viable! But invade they did in the ninth century, and the area became part of the Five Boroughs of the Danelaw (that is, Danish England) after the Treaty of Wedmore between Alfred, King of Wessex and Guthrum the Dane (*Alfred and Guthrum's Pact, 878*) setting out boundary territories as well as agreements on peaceful trade. Most of southern England split between the Danes and the English (Anglo-saxons) (Fig. 15).

Alone amongst English monarchs, Alfred is "The Great". At Christmas 877, he was a fugitive in the River Brue (Parrot) Riverscape, without army, land or wealth. In the spring of 878, in less than a month he was King of Wessex, having conquered the Danes at the Battle of Edington. He fortified settlements and reorganised rural life, including farming. He made Wessex a bright light of civilised life and learning with a European reputation. Also remarkable. One of his achievements was the double fort, a fort on each side of a river, so that—it was hoped—no Vikings or other invaders could ever again invade and conquer England by river.

A note from Sylvia Haslam:

When I started river research in the mid-1960s, pollution was considered more as unfortunate and better to try to hide it. So I was not told that the stretch of the River Rhee upstream of Barrington, dominated by the pondweed *Potamogeton penicillatus* (Fig. 16), was heavily polluted, although this only occurred where flow was held up—slow flow meant silt deposition, silt can incorporate much pollution, so the vegetation reflected this. Where water flowed freely, there was *Ranunculus* on open gravel. Pollution was much less as clean gravel holds little polluted silt. So the curly *Ranunculus* roots were able to grow in the surface gravel and anchor well. Diversity was low: *Ranunculus* was present (a good sign), but in good conditions there should have been about 8 aquatic species in a 20–25m length. There was not. Pollution here was properly assessed by the plants here.

Fig. 15. England at the time of Alfred the Great, showing the boundaries of the treaty between Alfred and the Viking leader, Guthrum, believed to have been signed at Wedmore in 878

As I gradually learnt more about what river plants tell us, I started to wonder, then to become sure, if this part of the River Rhee was indeed polluted. When I was sure enough to ask a specific question, it was acknowledged at once. There was no intention to deceive, merely it was one of those things you did not choose to talk about. As pollution became more monitored, and more illegal, it became better known. It can, though, still be hidden as much, or more, by ignorance as by intent.

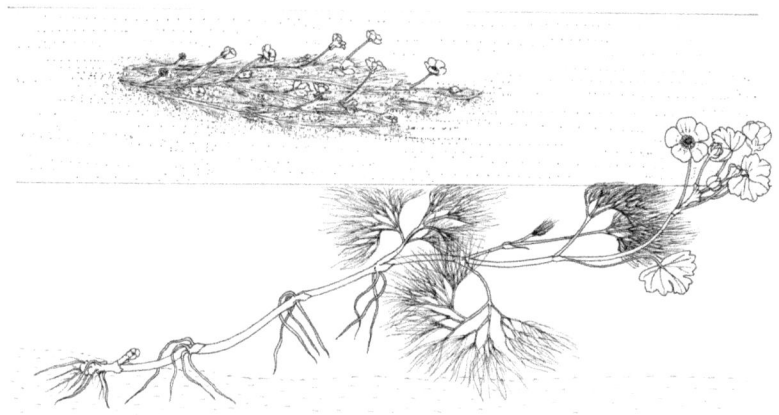

Fig. 16. Potamogeton penicillatus

So pollution started at the main source of the River Rhee where downstream water was moved upstream. Spring water has the lowest nutrients in this catchment, as it has not picked up much by way of ground surface chemicals. Downstream, chemicals *do* enter with run-off water, so nutrient (and probably pollution) status increases accordingly.

The River Rhee used to run through the bottom of the valley at Barrington, and the remains of the stream are still to be seen. An unusually long mill leat was built and is still maintained. This "leat" has now become the "River Rhee". The leat had to be maintained for the watermill to work, and for small boats to transport—so why bother about the other (real) stream, even though it was the old course of the *real* river? In fact, with the Drying Up of recent decades, there was less water.

Barrington also has a tributary running downhill from the village to the River Rhee. This was also downstream of the old village, channelled, so that transport (whether walking, riding or small boat) to the village centre was possible. (In 2000 there was still an old building sited by an old channel to the River Rhee—an ex-inn or trading house perhaps?)

A little downstream of Barrington, the next village is Harston sited on the slightly sloping bank on the south side of the River Rhee. Upstream, after a short, slower-flow length, it becomes partly silted. Silt, unlike gravel, is usually fairly rich in nutrients. Downstream there has been some movement and change in plant clumps in their placing, size and shape. Way back in 1970 there was almost-chalk vegetation in the river. Perhaps too much water starwort (*Callitriche* spp. Fig. 17). "Perhaps!" Plants are very sensitive to the habitat: but sensitive to all habitat factors, not just one! Upstream there is yellow water lily (*Nuphar lutea* Fig. 18, high nutrients, and strapweed (*Sparganium emersum*—*see* Fig. 14) suggesting, in a chalk river, downstream eutrophication, or raising of nutrient status. Here at Harston, this is hardly far enough downstream. Nutrient levels were up mainly because of the pollution

Fig. 17. Crowded Callitriche spp (Water Starwort)—too little water

Fig. 18. Nuphar lutea (Yellow water lily or Brandy-bottle)

described above, which in slow stretches is, as here, deposited, and so the water lily and strapweed can grow.

Around the millennium for about a decade the nutrient-rich siltier community crept downstream, under and past the road bridge, the water lily leaves being unusually small. In the downstream reach, the *Ranunculus* clumps became smaller with individual shoots becoming shorter, down from near-2m to barely 20cm, and with fewer shoots growing at the edges of the clumps. Drying Up means loss of water. Loss of flow means [polluted] silt which drops out of the water and is deposited, not sent downstream to join what is already the slow-flowing, Lower Cam and River Great Ouse. With several species involved, conclusions are near-certain. (If there is only one species, it is easier to ascribe changes to the wrong cause.)

Then came a flood-year. The water swirled, the bed was scoured, most of the silt on the top of the bed was washed out. The downstream water lily and strapweed vanished: they remained upstream, where now, after many years, they were well-rooted and established. *Ranunculus*, in contrast, grew greatly in the swifter (more oxygenated) flow, with a gravelly bed for root anchorage. And, of course, far more shoots, many reaching over 1m long. That is, the overall habitat can bear both the *Ranunculus* and the water lily/strapweed communities and the community can tip either way.

If the River Rhee at Harston had been almost unpolluted, there would be no water lily nor (probably) strapweed. They came because a polluted silt substrate allowed these to grow, and much-hindered the growth of *Ranunculus*, but this silt was easily washed out by more water and faster flow, so this fascinating change happened. A very nice example of reaction to a chemical change. If this upstream community had been in the Bourn Brook, with the appropriate other species, it would be ascribed to a clay outcrop. Difficult. The River Rhee, like the Bourn Brook, suffered badly in the Millennial collapse. (The Millennial Collapse was the surprising loss, *c.* 1995–2005, of both diversity and cover of aquatic plants in most of the more populous parts of Britain. It did not affect, for example, the far north west of Scotland.)

Upper River Cam and River Granta—usually labelled now as the River Cam (Fig. 19a, b, c)

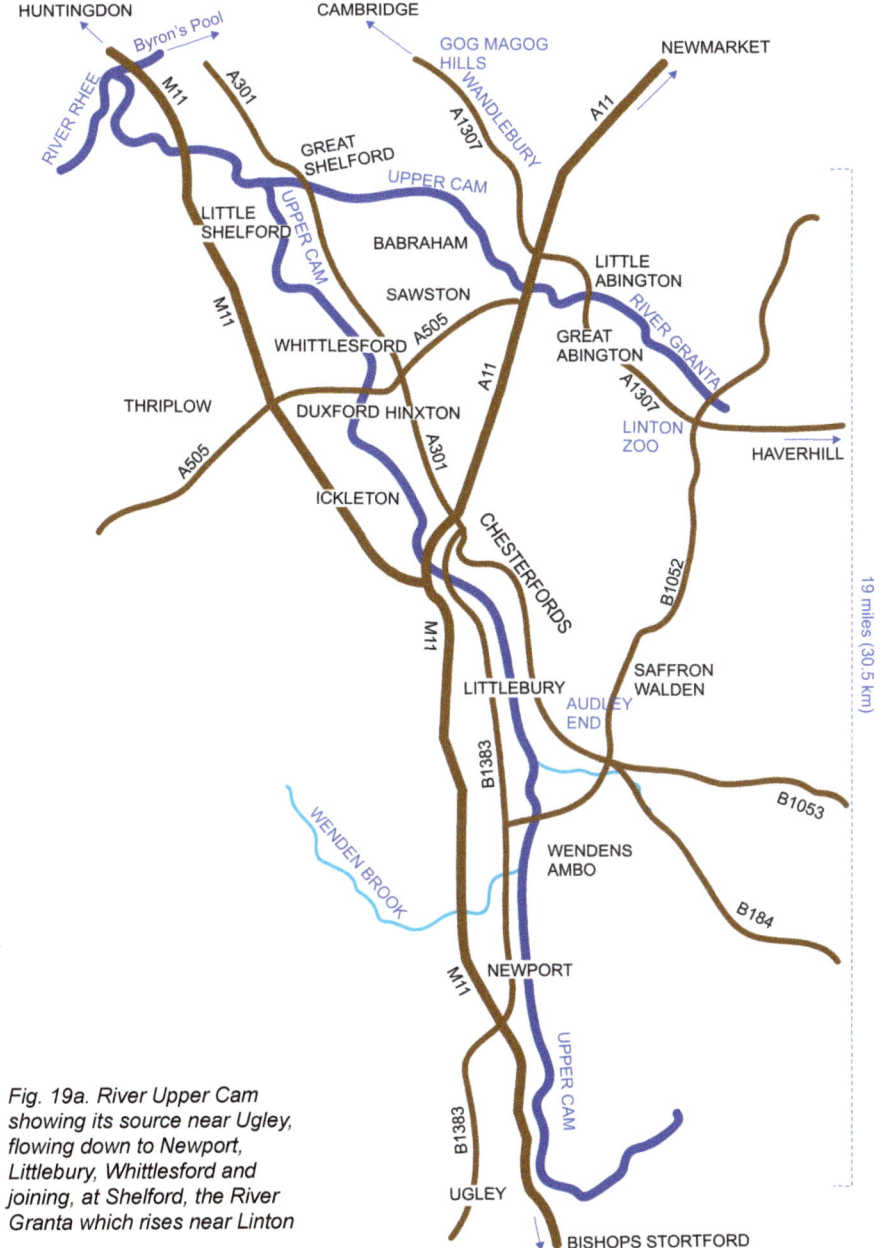

Fig. 19a. River Upper Cam showing its source near Ugley, flowing down to Newport, Littlebury, Whittlesford and joining, at Shelford, the River Granta which rises near Linton

Third around the arc, the Upper Cam is the longest of the four tributaries upstream of Cambridge, which had good quality wharfs (even at Saffron Walden). This tributary is now just a "rainwater stream" with much more water pre-land drainage—and any village, let alone a market town like Saffron Walden, must have been built beside enough water for its then needs (Fig. 19a). Figures 19b and 19c show river beds which have completely dried through over-abstraction (Source: *Cam Valley Forum Newsletter* "Let it Flow!").

Fig. 19b. River Granta's headwaters at Bartlow Barns August 2019 (Photograph courtesy of M. Foley)

Fig. 19c. Totally dry River Rhee bed under Sawston road bridge Stapleford, September 2019 (Photograph courtesy of R. Mungovan)

Saffron Walden is old (certainly Angle, possibly Roman). The Upper Cam and its long tributary the River Granta (fourth in the arc), is a mixed chalk-clay river with, of course, chalk and clay vegetation (Fig. 20).

There are no hills (not even almost-Downs as by the River Rhee) and no strongly chalk or, like the Bourn Brook, clay vegetation (Fig. 21).

The River Cam rises near the village of Ugley, near Stansted Mountfichet, increasingly known just as Stansted, after Stansted Airport (London's third airport). Confusion with other Stansteds is obviously less now. Both rivers have suitable (though decreasing) vegetation. Chalk and chalk-influence on the vegetation increases downstream.

Fig. 20. Typical aquatic vegetation in chalk-clay stream

Fig. 21. Typical aquatic vegetation in lowland clay streams

Upstream of Cambridge

The Upper Cam, comprising the Bourn Brook and rivers Rhee, Cam and Granta, upstream of where all join and become the (undoubted) River Cam, has many river villages with no large market square or civic building, but all are on a constituent stream of the River Cam, and getting their water from it. One, the Common Stream of Foxton, is artificial—a stream with no name—which flowed into the River Rhee. The Foxton Stream was historically a very important waterway but in 2025 it has mostly been piped underground, and dried. Other streams may have diversions from the main brook—for farm use—to take water to the farmhouse, farmyard, dairy, stables, to top up a duckpond or for drinking (now mostly "Ghost") ponds for plough horses at work in the fields, or for mills, for example, in Meldreth (Fig.22), Shepreth,

Fig. 22. River Mel, 2019: under the Mill house, showing the mill stream whose wheel has long gone, but still with the horizontal stanchions where it was once supported

Shelford, Grantchester and Barrington. Mills may be across the stream as at Shepreth, or be on a short or long diversion channel (leat) such as at Barrington.

There used to be many modifications for fishing, as shown for example in Figure 23, but they are less easy to see now. There may have been water meadows like those at Melbourn, Cambridgeshire, whose mills and associated flood plain are well documented down the ages, from the *Domesday book* through to the mid-twentieth century and present-day (Fig. 24). These were mostly active between the seventeenth and nineteenth centuries with (then)

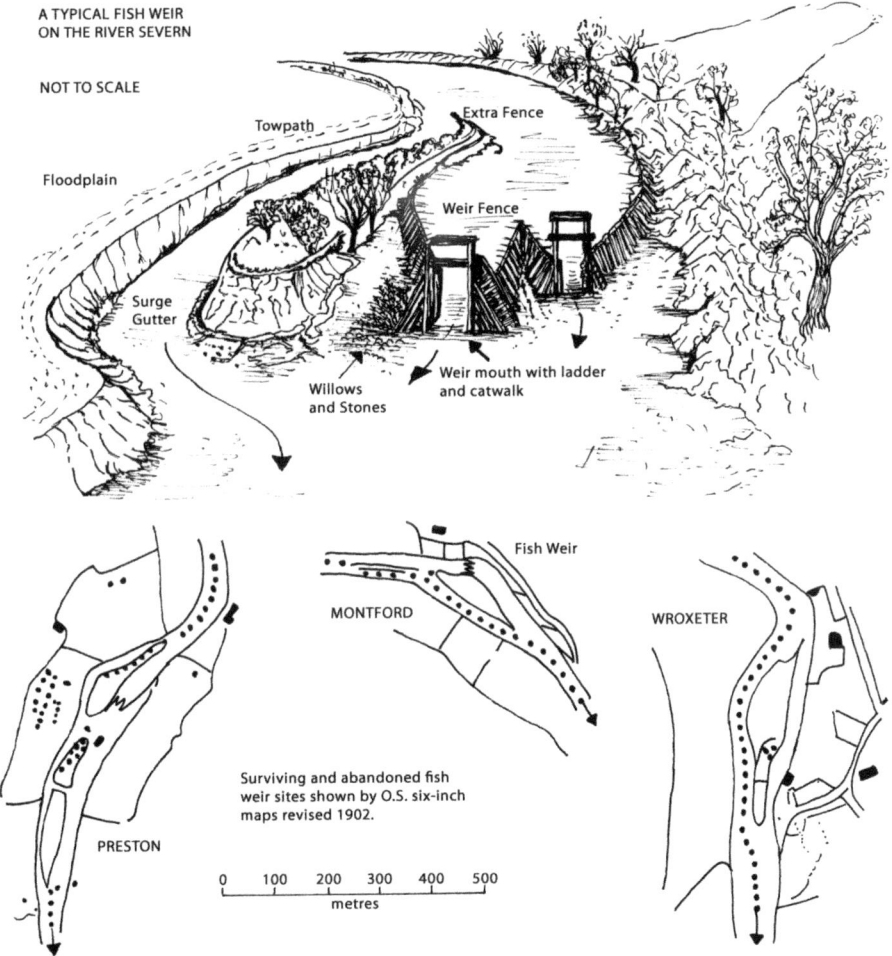

Fig. 23. Typical fish weirs on the River Severn and abandoned fish weir sites

lots of small channels criss-crossing the meadow, with hatches to control and guide the good, spring-fed chalk water. There were certainly both bridges and fords. There is the saying beloved of Oxbridge undergraduates for centuries that Cambridge is the better technical university because Oxford did not manage to build a bridge despite being older so more prestigious. Be that as

Fig. 24. River Darent, England. **A.** Near Dartford. **B.** South Darent. **C.** Horton Kirby. **D.** Farningham. **E.** Eynsford. **F.** Shoreham. **G.** Larger water meadow channels near Eynsford in 1867). **H.** Water meadow channels, only slightly decayed, Dorset. Adapted from Haslam, 1991

it may, there are, or were, plenty of fords, now tarmacked, over small streams with footbridges at the side.

The next development of course is a proper bridge. These were constructed over unpaved fords, with nothing, stepping stones, or a simple bridge at the side—later, bridged over (Fig. 25). Then there are abstraction bores and suchlike, where water is taken for supply even, for example, on the upper Bourn Brook. With the ever-increasing population and need for fresh water, the flow of water in chalk streams in particular is being depleted due to aquifer abstraction, as well as a significant deterioration in the quality of the water in recent times, as shown in Figure 26.

Fig. 25a. Stepping Stones

Fig. 25b. "Clapper" Bridge

Fig. 26. The ecology of England's chalk streams in 2019, showing the extent of chalk bedrock (outcrop) and whether each stream is of Good, Moderate, or Poor status. Also included is the number of streams flowing in each region. (Illustration compiled using resources from "The State of England's Chalk Streams" Report 2019, www.hyrdro-geology.co.uk)

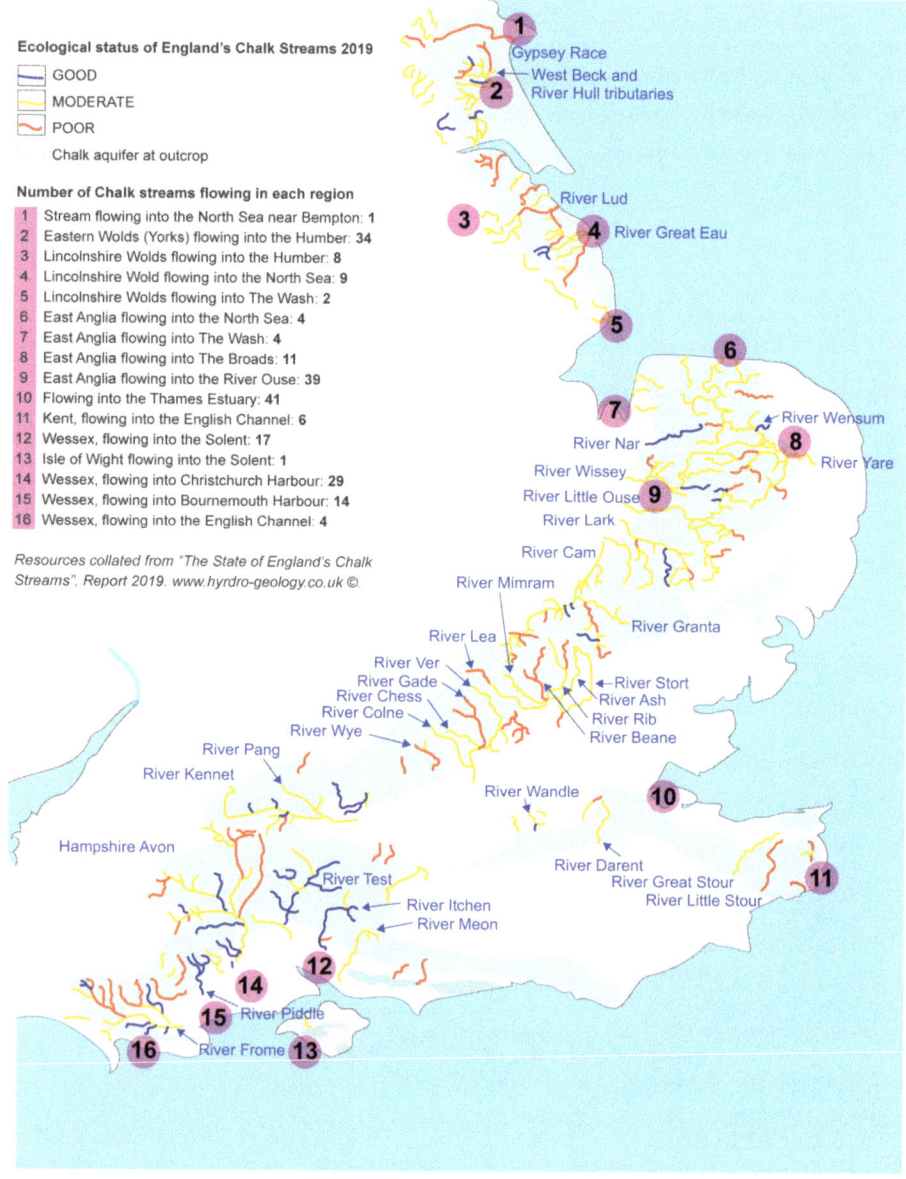

Coming down to Cambridge, as the streams coalesce, the river gets larger, and town-influence appears. For centuries, university undergraduates and townsfolk have used this and the stretch along the College Backs for recreation (including bathing, until into the twentieth century). Punts here, as at Oxford, were and continue to be favourite boats.

Cambridge

Next downstream is Cambridge, "Grantebrycge" of the Romans, which has all the proper things—a hill with a castle (Cambridgeshire County Council Offices from 1932 to 2021). The Romans built up the hill; the Anglo-Saxons by the river.

Fig. 27. Newnham Mill and Pond. (2025: the old mill is a restaurant called "Millworks" with inside and outside dining)

Fig. 28a. The commercial port of Cambridge c. 1780

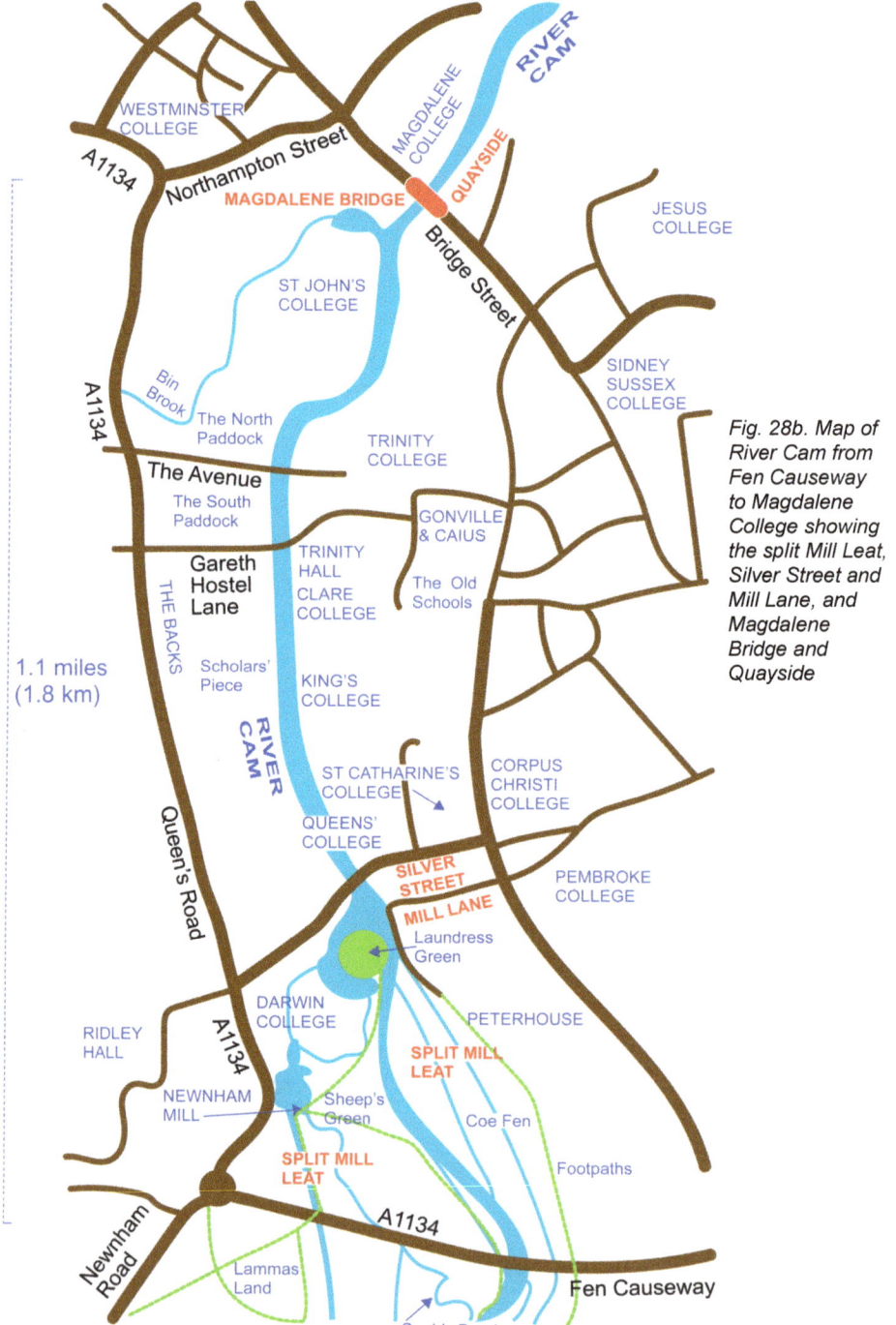

Fig. 28b. Map of River Cam from Fen Causeway to Magdalene College showing the split Mill Leat, Silver Street and Mill Lane, and Magdalene Bridge and Quayside

Shortly upstream, in the winding valley of Cambridge, lies the old Cambridge Mill in Newnham (Fig. 27). A leat of equal size, but straight, was later constructed flowing along the valley side, more directly into the town and rising enough to turn a wheel. This town mill area has the Mill Lane and old commercial riverside port of the town (Fig. 28a,b). The two channels join above Silver Street Bridge upstream of the "Backs". Magdalene Bridge now divides the upper from the lower town river (Fig. 29).

In the mid eighteenth century, Daniel Defoe wrote during his tour of the whole of the UK, "It is scarce possible to talk of anything in Cambridgeshire but Cambridge itself, whether it be that the county has so little worth speaking of in it, or that the town has so much, that I leave to others." Even now Cambridge dominates the county in a way that seldom happens elsewhere.

The town of Cambridge first comes to notice as a farm in 1500 BC. The first castle, on what is now Castle Hill, was built in the first century BC, presumably as a minor fort. This was rebuilt, bigger and better, by the Romans in *c*. 70 AD, and again in *c*. 410 AD. "Grantebrycge" (Cambridge) or Grauntcester (Grantchester, close by) was named by 500 AD. The Romans left and the Vikings arrived by 875. William the Conqueror rebuilt the castle, and the town was Chartered in 1120.

Fig. 29. The impressive Magdalene Bridge, built in 1823, with Magdalene College in the background

Time after time what remains of a river town, whether Florence, Megéve, Frankfurt, Norwich, Nuremberg, or York, and many more, follows this pattern. Anglo-Saxon towns had the town developing on a tributary to the main river: the river is there for transport, but in ordinary conditions the houses were not flooded as they were on the small stream. The tributary turned a mill and supplied water to the town, and its public buildings, including the Church. There would be a sub-adequate wharf—boats being used for transport and probably for fishing with nets, as well as a fort of some kind, for protection. Alfred-the-Great developed the double-fort, one each side of the river, for protection, though usually (until much later) most of the development was only on one side of the river.

In the next stage, the town looked more as we would expect! The market place, plus civic building, plus large church or cathedral became well established and usually remained unless or until overtaken by development. All increased in size and grandeur. Bridges appeared, maybe with tolls, maybe donated by a local Great Man (cleric or noble). Fortifications appeared, but varied with conditions. There were more in Europe than in England. There tended to be more where Danes and other Vikings invaded up rivers, and the local people defended themselves. The wharfs and other commercial development increased in size, technology, and complexity and there may have been development for a particular purpose, for example, wool, textiles, foreign trade, metal work, regional capital, and so on. Usually these evolved naturally from the previous stage.

Not in Cambridge, however! In 1445 King Henry VI decided to place his two "royal and religious" institutions one in Cambridge (Kings College), the other in Eton (Eton College) on the Thames alongside Windsor Castle—the premier palace fortress of the realm. A great bite was taken out of the riverside area (modern-day "Backs" of Cambridge). The main wharfs and perhaps harbour were completely gone. The main church, Great St Mary's, remained but the Old Market Place between the church and the River Cam had gone, and the Market Place appeared on the other side of the church. Figure 30 shows a c. 1350s Cambridge, with the slaughterer's diversion stream—King's Ditch—encircling much of the town, and plenty of new channels. As a town develops, so it needs more good water.

So, a thriving trading, fishing and agricultural town was deformed and never recovered. Instead, one of England's architectural glories appeared: King's College, with its fan-vaulted, unique chapel. For some reason, the missing commercial port was not replaced. When did the University start? The year 1209 is cited, when a few boys accompanied a clergyman from Oxford, to be taught as they wanted. But...? Anyway, between 1209 and 1284. The University also is one of England's "glories". There appears to have been a vigorous rivalry between town and gown over the centuries and the wearing of a university gown was a much-coveted privilege.

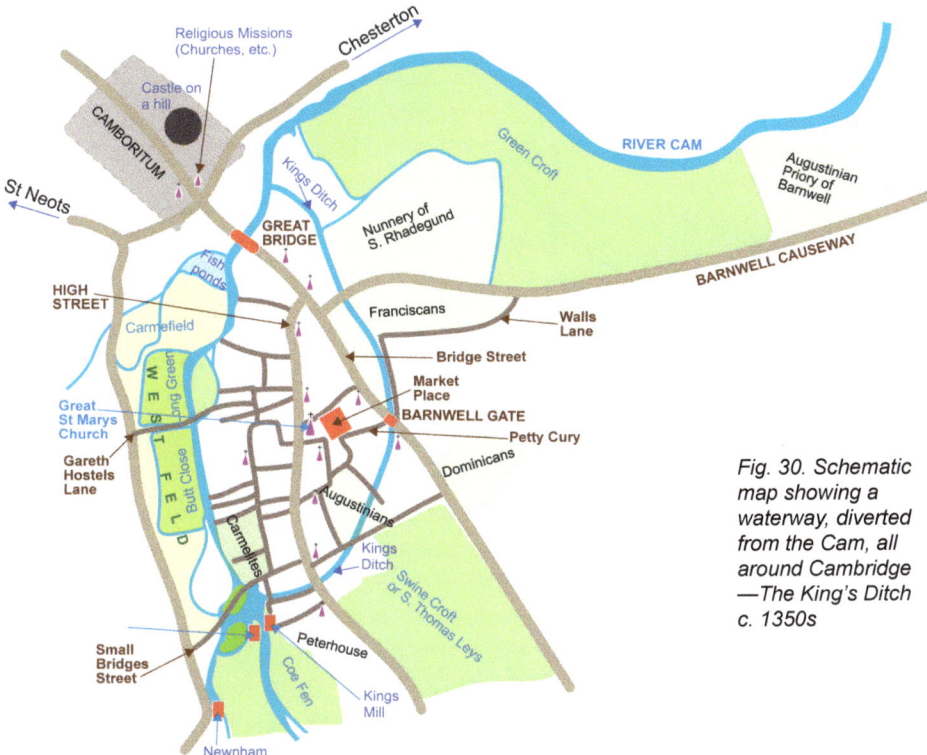

Fig. 30. Schematic map showing a waterway, diverted from the Cam, all around Cambridge —The King's Ditch c. 1350s

Magdalene College, a Church foundation, was concerned about the purity of its water for the health of its undergraduates and in mediaeval times diverted a stream to bring good water from the south. This was forgotten. When mains water was brought to the area in the nineteenth century, Trinity College Great Court fountain was left with the old supply. This also was effectively forgotten, until Churchill College, when being built in the 1970s had its cellars flooded. Even later (1990s), a friend dug a small (c. 3m × 1.5m) pond in her garden nearby and Great Crested Newts appeared (Fig. 31). The only explanation found was that this same diversion of the water for the Colleges had probably harboured this rarity for centuries.

41

Fig. 31. Great Crested Newts (Triturus cristatus): top centre—female (can reach up to 17cm, but usually between 11cm and 13cm), male, lower right, in breeding colours; juvenile, lower left, normal dark grey–black colour, amongst Potamogeton natans (Pondweed)

"Hobson's Choice" was a well-known saying meaning there was no choice. Whilst in most such stables a rider could pick the horse he wanted, Mr Hobson insisted his horses needed proper rest, and would only hire them in rotation ("That horse, or none!"). Mr Hobson also brought (again!) clean water to Cambridge with a diversion from a source at Nine Wells Springs, near Great Shelford, to central Cambridge with a fine fountain open to all in the Market Place (now moved to the side of the centre). Here the conduit water split into two to feed the fountain, and to supply Christ's College (Fig. 32). In 1855, a Gothic Revival gabled fountain was erected, in place of the conduit, but this was demolished in 1953. Its water-flow was finally cut off in 1960 during the construction of the Lion Yard shopping centre.

Hobson's Conduit, from Hobson's Brook, also provided water to clean the (then) main streets, with runnels of flowing water on each side (Fig. 33).

By the 1970s over-abstraction and the 1976 drought brought more Drying Up, and the springs at Nine Wells almost ceased to flow. Water was pumped from Babraham downstream (environmentally-minded for the time) so that

Fig. 32. Hobson's Market Place Fountain: From 1614 the conduit brought fresh water to the Market Place. In 1849 (Victorian), damage to the Market Square by fire resulted in the original conduit being moved to its present position at the Lensfield Road-Trumpington Street corner

Fig. 33a (left). December 2023. Hobson's Conduit at the junction of Trumpington Street and Lensfield Road.
Fig. 33b (above). December 2023. Conduit runnels with flowing water down Trumpington Street, Cambridge

the water from Nine Wells was again full and flowing through the conduit and city runnels. Natural spring water has less nutrients and a different composition to that entering downstream with run-off from fields and gardens. At first the surroundings kept enough of the original water to maintain its rare invertebrate community, but long-term they have not survived. So pollution started at the main source of Hobson's Brook. And now the Nine Wells, once a Site of Special Scientific Interest (SSSI), has been reduced to the level of Local Nature Reserve (2020). This was done by those understanding water quantity and heritage, but not QUALITY — spring water has less nutrients and a unique temperature.

Finally, as Cambridge is an Ancient University, it has the advantage of being studied by many eminent scientists over many centuries, and a few of these have been interested in the River Cam. As described in the first book in this series (*Drying Up*), plant species have uniquely in this county been recorded intermittently over three and a half centuries from *c*. 1660. However, and regrettably, two-thirds of the species recorded earlier have been lost. Figure 34a,b,c shows the distribution of aquatic plants on the Upper River Cam for the years 1971–1972, 1976 and 1978–1979.

One historical major event Cambridge no longer has is Stourbridge Common Fair, the last of which was held in 1933. Daniel Defoe, novelist, traveller and government agent, toured "*The Whole Island of Great Britain*" (1724) and visited the Fair at Cambridge, describing it as:

> *12 square mile, crammed with merchants, dealers, buyers* [often having £10,000 of orders, £10,000 in 1700!]. *All trades that can be named in London can be bought at the Fair, e.g. food and drink, hops (major), wool, wrought iron and all metal. Goods could travel to London by road* [waggons], *or water* [via the River Cam, Kings Lynn and River Thames]. *Trade was with, among others, Leeds, Rochdale, Bury (Lancashire), Manchester, Exeter, Taunton, Bristol, Birmingham, Sheffield, Surrey, Kent: all over the country* [even more than a pop festival or rave, these days!].

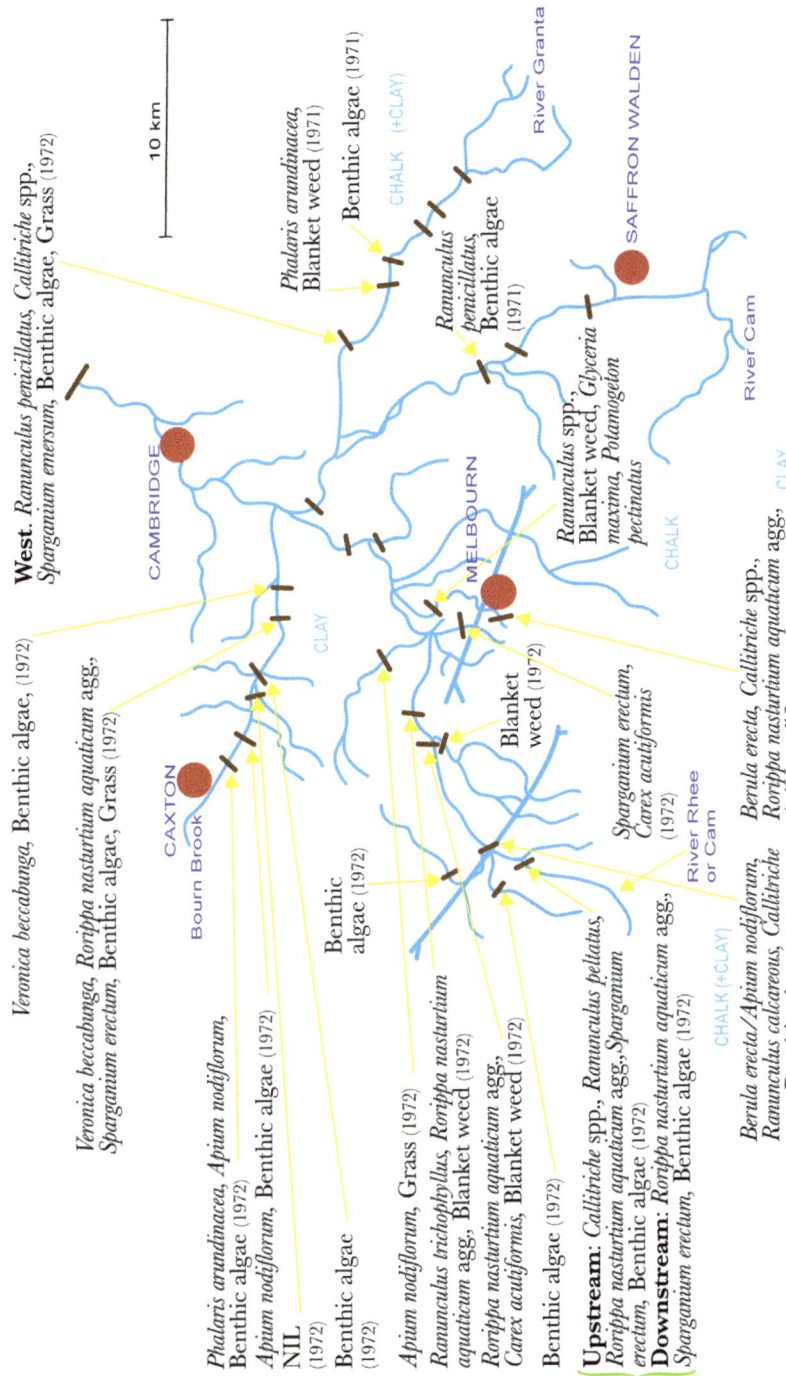

Fig. 34a. Aquatic vegetation recorded in the Upper River Cam 1971–1972

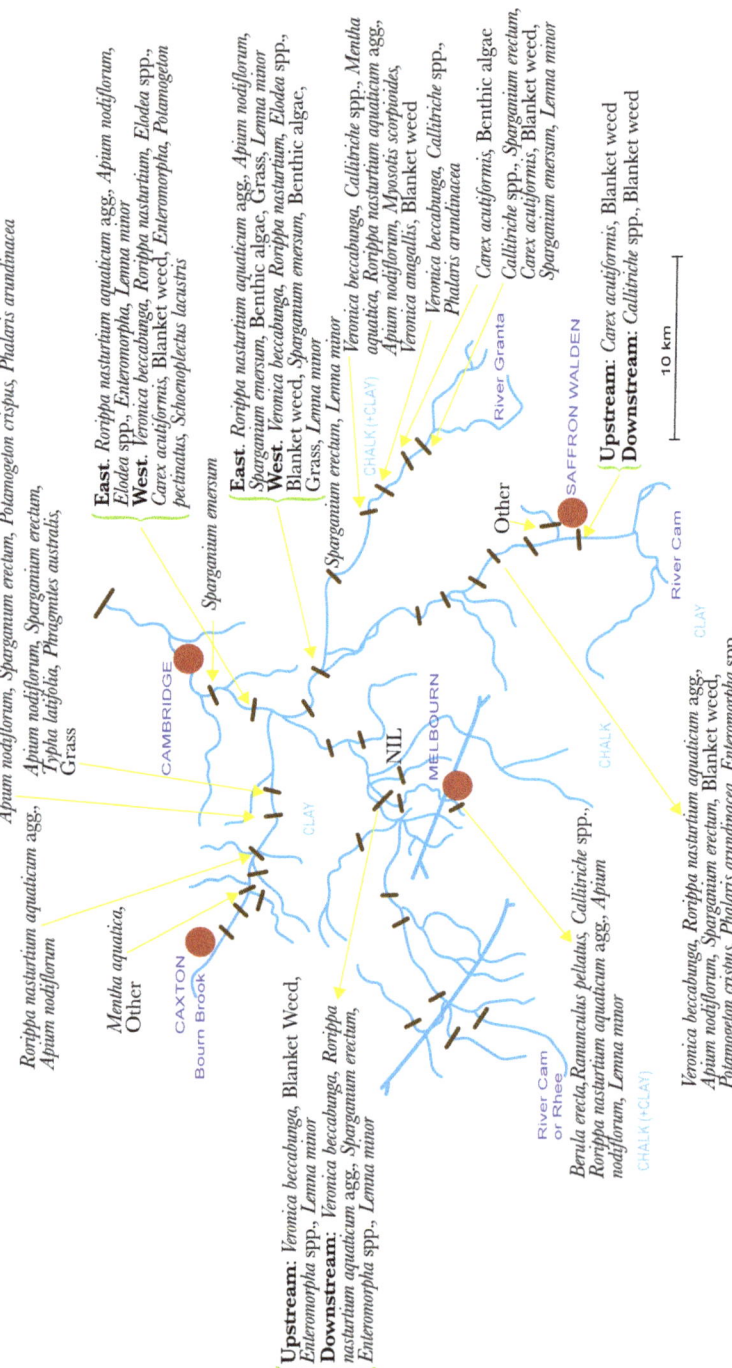

Fig. 34b. Aquatic vegetation recorded in the Upper River Cam in 1976

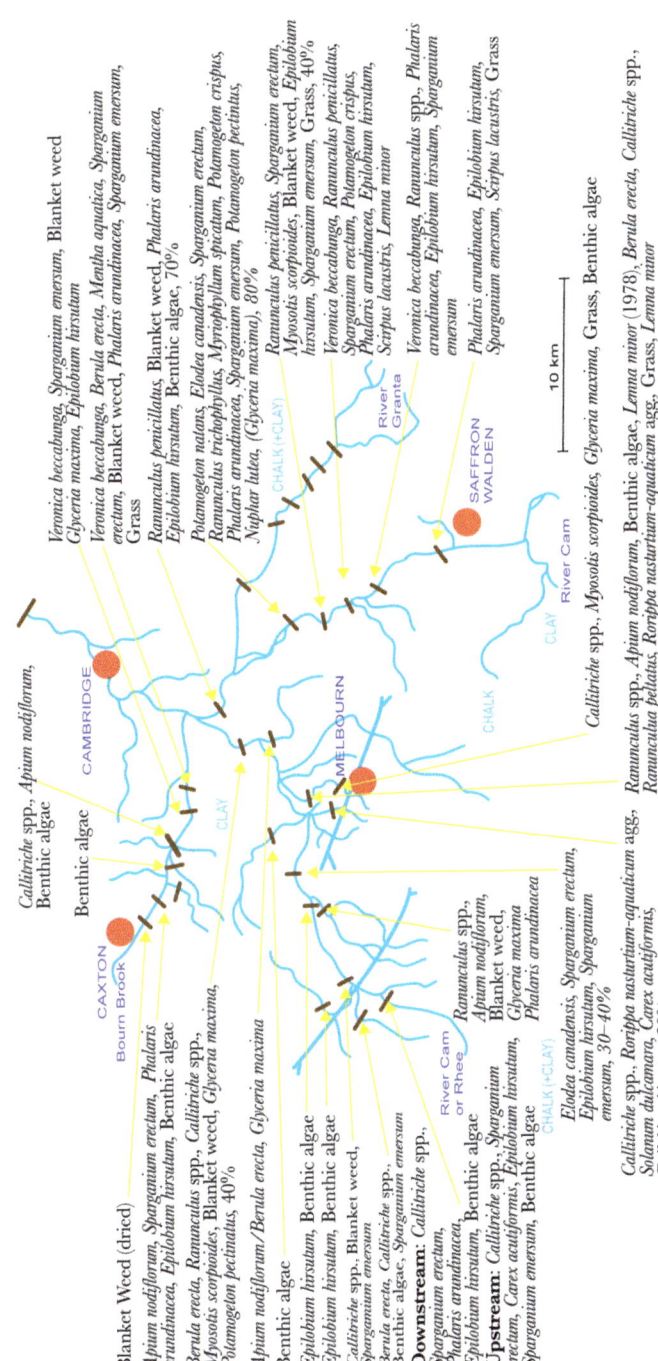

Fig. 34c. Aquatic vegetation recorded in the Upper River Cam in 1979

Place Names

In the Cam catchment, place names—as is usual for England—go back to Roman times (roughly 50BC–400AD) maybe with a hint of Celtic in the older ones. "Grauntcester", of course, is Cambridge (or Grantchester, a little upstream of it). Many are Anglo-Saxon, but names are scattered through history right up to the present. For example, Bar Hill and Northstowe in Cambridgeshire are late twentieth-century villages. The names have a chequered history, having had to put up with many influences: Roman, early, mid and late Anglo-Saxon, Norman and Tudor (Medieval), Stuart, Georgian, Victorian and recent.

Standardisation came *c.* 1815 after the Napoleonic Wars, when the British government (ever-ready to win the last war but one!) embarked on a major mapping exercise, firstly to find out exactly where there were stores of ammunition (ordnance) and then what the communications would be by road or water for the military. The Ordnance Survey (OS) has been with us since: a highly respected organisation. In the 1980s, the best geological maps to be found of Belgium (battles having occurred there) were the First World War OS ones.

Standardisation was not always perfect though. An imaginary conversation deduced to be often repeated, was:

> *Inspector:* "What is that called?"
>
> *Countryman:* "That's the river [*sotto voce* (whisper)], you are stupid. Obviously it's a river".

And so the nearby river was named "River". This led to the large numbers of present rivers called, for example, Avon, Ouse and Derwent—all these names mean "River" in various tongues!

Far too many Cam catchment river-names unfortunately just come from a—usually Anglo-Saxon—person or group of people such as Barrington, Bera's ton, or village such as Mepal, Meothas' people. Around 10% do have some sort of river name. The main groups are:

> (1) **Boats**. Water transport can be major for a place, such as Littleport downstream of Ely, or can be just a landing place such as Histon, or Meothwold, which has a tythe. (Also "tithe"—being one tenth of annual produce or earnings, formerly taken as a tax for the support of the Church and its clergy.)

(2) **Ford or Wade**. Where you can walk or ride across the river bed. There are plenty, for example Kentford, Mundford, Lackford.

(3) **Bridge**. Only a few: Bridgham, Cambridge.

(4) **Mills**. The Rivers Mel and Mild. In Anglo-Saxon times most were corn mills. Diversification of function came with the explosion of technology when the workforce was much reduced after the Black Death in the 1340s. The River Rhee has Melbourn and Meldreth; the River Lark, Mildenhall and Barton Mills. Most mills, have not given their name to their river.

(5) **Produce.** Eels, of course: Ely (eels were often used as money); trout (Farnham); dairy (Wicken); alders (Aldreth, for clogs); leeks (Lackford); ash (Badwell, for wood); flax (Linton, for linen. Flax needs to be retted—soaked—before being made into linen).

Around another 10% of settlements have names with other water descriptions:

(1) **Flowing waters**, often "wells". Because of so much recent Drying Up, we now think of wells as still waters whilst the former representative "well" had water welling up. There are quite a lot, for instance: Feltwell, Holywell, Burwell (well by a fort), and Ordwell (by Ord River).

(2) **Still waters**, for example, Lynn (pool), Livermore.

(3) **Islands** in the deep fenland, in meres and rivers, for example, Ea, Ey, Eye, Ile, Isle.

River Settlements

River settlements typically started as single farmsteads or mills. A few large ones were planned in advance so were originally built up as towns. Then the farmstead was joined by more, and a village, then a town, resulted. A sizeable village would diversify with farms, carpenters, thatchers, bakers, fishermen, carters, and so on, accumulating basic trades and becoming near-independent units.

What is done with water depends on its availability, the available technology, the ideas of the inhabitants, the superfluous wealth and the motivation to use these. Roman Britain had all in plenty, but the Anglo-Saxons who followed them, and saw them in action (so could have copied) had, for reasons we cannot understand, no interest. Roman conditions were kept to some extent by monasteries and convents: but civil society lost interest. Comparisons are difficult, but Britain barely reached Roman levels of services such as water supply and domestic heating much before 1900. Figure 35 shows a variety of historic river villages similar to those of the Cam catchment.

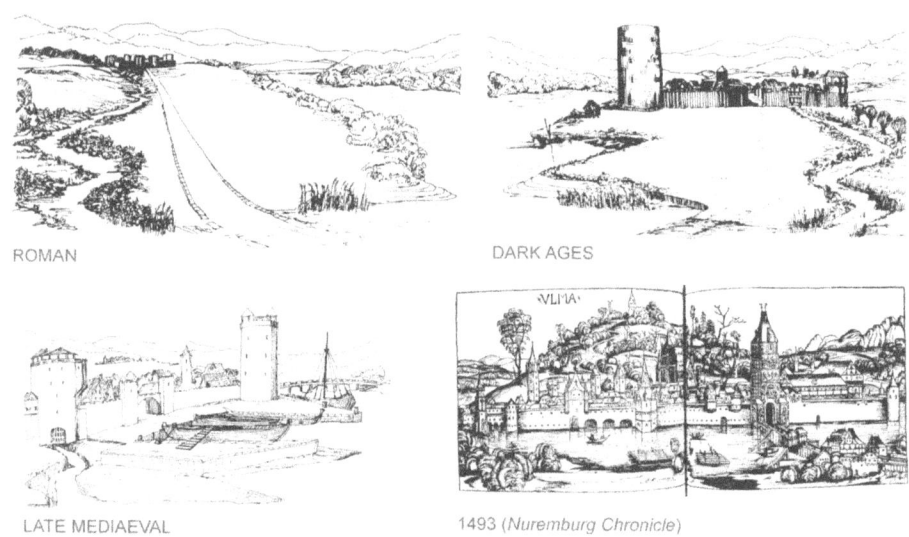

Fig. 35. Early stages in settlement development. (From S.M. Haslam, 1997. The River Scene— Ecology and cultural heritage. Illustrations by Y Bower/P.A. Wolseley)

Water is a pre-requisite. It is necessary, and it is heavy. In a wet country like Britain it is a waste of resources to spend hours daily carrying water home. It is better to build near water, or bring the water to the houses.

Villages may be at the sources (springs, rills) of brooks, by parks and meres, on stream banks out of flood reach, on stream banks within flood reach (probably trading), on small channels diverted from brooks or rivers (normally un-flooded), and so on, with many patterns and permutations. Fortified keeps and castles had water from underground streams via diversions and deep wells which potential besiegers could not easily access. Livestock drink, and farms were laid out so that cows and other livestock could drink close to natural sources—most fields had natural ponds now mostly Dried Up. Watermills needed strong enough flows to turn mill wheels. Fish needed waters of appropriate size. Boats were commercially effective in surprisingly small brooks (say, 75cm wide, 30cm deep) as well as where they were large enough to travel to, for example, The Hanse (German Hanseatic League area which, between the thirteenth and fifteenth centuries, encompassed nearly 200 settlements across eight modern-day countries from Estonia in the north and east, to The Netherlands in the west, and extended inland as far as Cologne), to Scandinavia, or to the Mediterranean.

Burwell, Cambridgeshire, specialised in boat-building. Reach had annual fairs with many visiting merchants, even from London, Venice and beyond. Other settlements specialised in fish, textiles, vegetables, a smithy, an abbey, and many more, though most villages were general farming communities (some to grain, or grazing sheep, cattle or geese).

A common position was to have the farms and village houses a little beyond the stream, where the land rises. Land near the stream was liable to flood and was managed as flood meadow: nutrient-rich, silty floodwater fertilising the grass, or water meadow: a complex water system of flooding, silting, fishing and grazing, constructed mostly during the seventeenth to nineteenth centuries. The villagers would then come down to the stream for fish and boats (door-to-door from Germany to Burwell). Where housing was on river edges, damaging flood was more likely, but navigation reached right up to the housing. The flatter the land the more it was prone to flooding. In the hills, flood plains tended to be very small, or local.

Figure 36 shows a farmhouse as indicated on a 1920s map in Comberton, Cambridgeshire, near a stream (*see* Fig. 13). Diversions from Tit Brook meant that water was available wherever it was wanted on the fields and in farm buildings. But did it have piped water indoors? Perhaps; perhaps not. It certainly did not have hot piped water downstairs and up at the turn of a tap! Did it have piped water for outdoor purposes: the small channels may be blocked, or dry up—or even flood?

Fig. 36. Water from Tit Brook a tiny stream running south into Bourn Brook was diverted to the farmhouse in the 1920s. In 2025, Tit Brook is still running, but the diverted channel is dry

The larger (and Upper Cam) villages are—as any in the ancient English countryside—a mixed bunch. The main road may be beside the river, at right angles to it, a little further off (when the land by the river is over-liable to flooding), or placed for quite other reasons. Most villages had duck ponds which were important not just for growing ducks, but also for watering horses (transport, plough) and other livestock. As Gilbert White records in *The Natural History of Selborne* "Cows could enjoy being deep in water in hot weather".

Settlement avoided the wet fens, obviously: so farms or other homes on the (now) wetland were built more recently. Old settlements on islands tended to own "summer land", that is, dry land on the slope down that was dry enough for crops or good grass in summer. Fenmen had access to the wet Fen, of course, from whence came peat (for heat and light), fish, reeds and sedge (for thatch, animal bedding, furniture, etc.). By the nineteenth century, concerned visitors were reporting squalid conditions and lifestyles, and there was much malarial-type fever. A reliable test for malaria was not developed until after drainage of the fens and improved social conditions had effectively eliminated this illness.

Different observers reported differently on the health of the Fen folk, some describing "pale, weakly people", others, a healthy population. (Further data is needed.) Anyway, whilst murder and mayhem are frequently recorded in the Somerset Levels (fenland), that is not so in the East Anglian Fenland: fewer recorders? Less friction? Or…?

Downstream of Denver Sluice the ground slowly becomes drier (that is, above instead of below sea level) and agriculture and villages become typical lowland as the river approaches the coast at Kings Lynn. A final twist to the tale is that, from the end of the last Ice Age (*c.* 8000 BC) to the thirteenth century, the Peterborough-Wisbech-Wash river carried Cam-Great Ouse water, then a severe storm altered the course so that it silted up. This meant many changes, including the loss of Wisbech as a major port, though here, by the river, the Dutch influence is still unmistakable! Figure 37, reproduced from *Dugdale's History of Imbanking and Draining*, Fenland Notes and Queries, July 1892, shows The Fenland before being drained *c.* 1652. (*Note the map has been drawn upside down!*) The Great Ouse used to lead (for the Hanseatic League towns especially) to most of the east midlands of England, now adding the Cam basin.

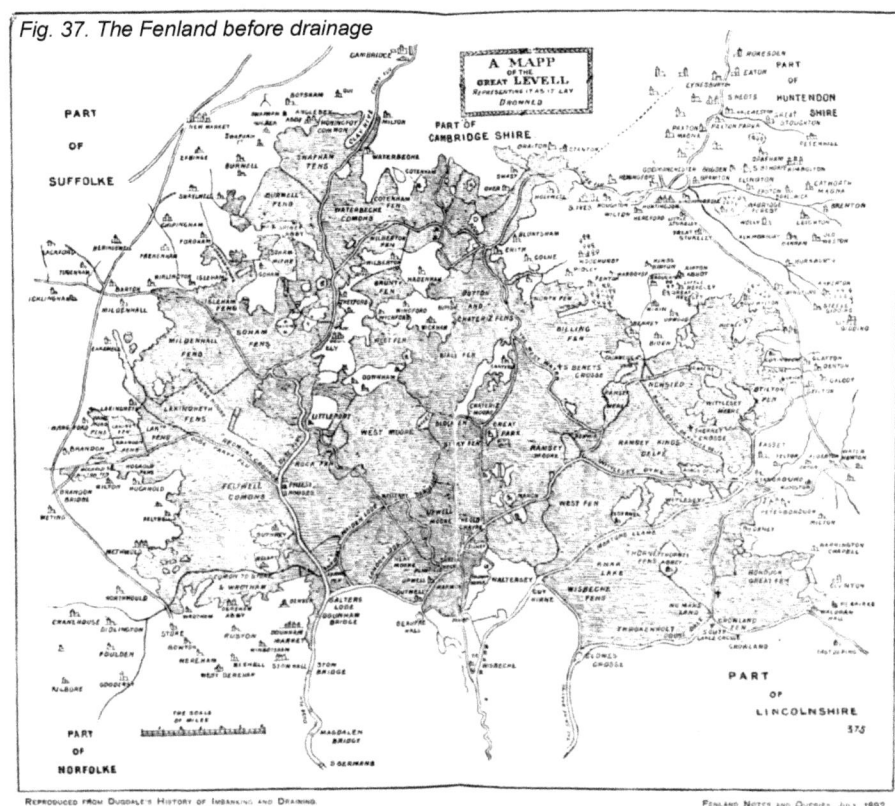

Fig. 37. The Fenland before drainage

Uses of rivers

Water

Water itself, of course, was the primary use of the river: water for drinking, cleaning, manufacture, transport, and to water farm crops and animals (Fig. 38). Fields and gardens needed watering in dry weather. Livestock did, *all* the time.

Fishing

Fishing had to and could be lucrative, and this was aided by the constructions for mills. Species could be anything from white-clawed (native) crayfish to minnow to salmon, and fry—shoals of it (see Daniel Defoe's *Tour*).

Fig. 38. Various uses for water taken directly from Rivers. a) Drinking—human and livestock. b) Laundry. c) Water for irrigation. d) livestock drinking directly from the river—special fenced area. e) Eighteenth century canalised river diversion channels for transport. f) Diversions for industry— here an old tannery and a mill with millers cottage adjacent. g) Navigation and transport. Illustrations by Y Bower/P.A. Wolseley, except "c"

Milling

Water mills (whereby power is generated) were invented in BC in the Mediterranean, and quickly spread. They were in Britain before the Norman Conquest in 1066. Corn mills squashed grains to a powder to be used for making bread. (Windmills arrived soon after the Conquest.) By *c.* 1500 most lowland villages had a watermill. Power was generated by moving water to turn a wheel (*see* Fig. 7). Other uses or constructions such as fish ponds, farms, fords, and bridges could be added.

Fish Ponds

Fish ponds ("lakes") and early moats could be managed specifically for different types of fish. In mediaeval times, the Masters of the Fishponds of a manor had the same status as the Masters of the Horse. There is little point having a small army without the food to support it! Ponds belonged to the

Church, to the Landlord, and to the village, the latter typically being downstream (so the fish could feed on the waste—maybe including sewage—but the people could drink clean water from upstream.

Transport

Rivers and brooks were main highways. Villages had small, everyday wharfs which now too often have vanished. There may have been washing stones where women washed, stepping stones to cross or walk along the stream, perhaps paths beside the stream for towing—though before canals, the animals or men are more likely to be walking over unenclosed land. Canals and River Navigations had money to invest in firm, safe tow-paths, along which horses could move quickly, making freight voyages much quicker—and shorter—since canals were made as straight as possible.

Below is an excerpt from a dissertation on Cambridge and its economic region, by John S. Lee, written in December 2000:

> Where possible, water transport was used to reduce carriage costs. Sedge could be shipped onto the River Cam from Reach via Reach Lode and from Lakenheath via the Little Ouse. In the eighteenth century, sedge was unloaded by barge at the quay by Magdalene Bridge, and the same practice probably operated three hundred years earlier: 3s 4d was paid to carry 1,000 sedges from the Great Bridge in Cambridge to King's College in 1467, and 200 sedges were purchased for King's College from the "Seggeshith" in 1451.

Where a river had been moved, the bed of the old course was typically inorganic (particles brought down from the lowlands above). Where peat around had disintegrated, the old course would disintegrate less, and the river bed sticking up was named a "rodden".

Recreation

Boys used rivers for sport, fishing, bathing and so on; more modestly, girls were more circumspect. (The famous scene in the television adaptation of Jane Austen's *Pride and Prejudice* where Mr Darcy plunges into a weedy lake is in fact un-historical. Mr Darcy, educated and country-bred, would not have risked death by swimming alone among water weeds—the literature is full of such drownings. Down the centuries, prominent—and therefore also less prominent—English and Scots, Welsh and Irish, have enjoyed leisure activities by, on, and in streams.

Legal Navigation

The River Cam was a Legal Navigation, so was made as straight as possible, with a strong tow-path. The same applied to the Great Ouse in this region. The main drains could also carry ships, and the smaller dykes, boats. These were mostly for local traffic to villages and outlying farms which, pre-Enclosure Acts, were few.

Water Fowl

> Oh what have you got for dinner Mrs Bond?
> There's beef in the larder *[mens' work]* and ducks in the pond
> > *[women's work].*
> Dilly dilly, dilly dilly, come and be killed,
> For you must be stuffed and my customers be filled."

(Old English Nursery Poem, see Citations).

Duckponds could be—as in the poem—for semi-tamed and well-fed duck which lived there. Mallard, coot, moorhen and many more. Swans, if arriving, were a different matter as they belonged to the sovereign, and were [officially] under Government control. Depending on the habitat, the duck would nest on the fringes, or in specially made duck houses. Duck were in abundance and, in the country, could be an important source of food. By the eighteenth century, duck decoy ponds were developed with enough wild duck to start depleting the population. These might be constructed, or use part or all of an existing pond. "Tame" ducks attracted wild ones. There were inlets, often reedy-fringed, and dogs chased the ducks into the far end of the inlets where they were caught, the "tame" ones were released back into the pond and the wild ones were sold—including to markets and shops in towns.

Thatch

In the lowlands, thatch was often grain stems, so in England it was predominantly wheat straw, though also other grains and any reeds or sedges. Wheat-thatch likewise is done better and lasts longer (25+ years) for the manor or Church. However, in The Fenland itself, **thatch** was usually reeds or sedges, done expensively to last 60+ years on the main buildings, and cheaply on out-buildings.

Local Enterprise

Local Enterprises were likely to use water, for example, timber, withies, bedding, irrigation, thatcher, butcher, smith, baker, domestic animals, including horses, who drink water.

River towns

Five river towns can be recognised on the River Cam (plus the Lower Great Ouse). Upstream is Saffron Walden, a small market town. Then the County Town of Cambridge (which also is Roman). Next comes Ely where in Anglo-Saxon times the abbey and cathedral were founded (so Ely, not Cambridge, is the present-day site of the See (diocese)). Only a little downstream of Ely is the second small town of Littleport—a very useful sub-port for Ely. Finally is the port of Lynn, existing but little-known to historians in pre-Roman times. A Lynn is an estuarine pool—which indeed Lynn, South Lynn and Bishop's Lynn all had until development of port and industry removed the pool water (as it has at so many others), and especially after the thirteenth-century diversions, commerce became very important. Norfolk, like Somerset, became immensely wealthy (for the times) from sheep.

"Walden" was to do with woods, and Saffron is the pink Saffron Crocus (*Crocus sativus*, Fig. 39), with its vivid yellow stamens, much prized, especially in Roman times, for its culinary and dyeing properties.

Fig. 39. Saffron Crocus (Crocus sativus)

This town marked the end of navigation for most boats (goods going further upstream had to be transferred to smaller and narrower boats. Commercial boats continued to take goods up and downstream as far up as possible: often as high as the stream was wide enough for a rowing boat, and deep enough for it to be afloat (about 1m). A man or a horse could pull it (Fig. 40). Alternatively (mid to late seventeenth century), as shown in Figure 41,where the water was shallow and a hard river bed present, carts and riders could travel along the bed of the stream. (Roads, remember, were pretty awful, and this was well before tarmac (see Daniel Defoe's, *Tour*).

Fig. 40. Early nineteenth century river and horse-drawn boat

Anyway, Saffron Walden was the head of easy navigation. There is often confusion about the meaning of "Navigation", and books are confused too. Does the word mean where sea-going boats reach? When there is much commercial shipping? Where there are only boats carrying goods for a particular farm or manor house? In fact, none of these! Any stream which can take boats or people is navigable, but the (mostly early eighteenth-century) Navigations were where one or more people got an Act of Parliament, made a stretch of river truly and easily navigable (or canalised), and charged fees for boats to travel along it. The River Cam in 1780 was, in this sense, navigable for large boats and much freight up to Cambridge.

Fig. 41. Shallow river beds were regularly used in the mid to late seventeenth century as roads for carts, people and livestock

There is even more variation in where boats landed. Boat transport in general was not only cheaper, but quicker for freight up to the nineteenth century, though liable to delays from flood and drought. There were major river ports, minor river ones, landing places for farmhouses, for villages: everything from Kings Lynn to where a house kept (or indeed may still keep) a little rowboat for recreation where its garden met the river. Famously, and well-recorded, the Vikings came up the River Cam, then the River Rhee and invaded that area—and no doubt elsewhere. Ultimately the Cam catchment fell within the Danelaw (*see* Fig. 15). (The English, to the mystification of continentals even now, still know the Danish, pre-metric peculiar numbering systems of 12 inches in one foot, three feet in a yard, and 1,760 yards or eight furlongs in a mile.)

Sea transport linked England with the continent and beyond. The Romans built many sea ports, such as the garrison town and major port of Dover, and lesser ones down the east coast. However, a fascinating television programme called *The Grain Run*, narrated by Pete Morgan (BBC 1985), travelled the waterways used by the Romans, where they still exist, to reach their outposts in the North, via the River Trent. Ships collected the rich grain from the agricultural lands of the River Cam, took it up north via rivers and canals to Torksey on the River Trent, which town is reported in *White's Directory* as being the Roman Station, *Tiovulfingacester*, built at the entrance of the Fossdyke Canal to secure navigation and as a storehouse for grain. The Fossdyke Navigation dates back to Roman times, giving it a strong claim to be Britain's oldest canal. This seems an unnecessarily long way to the sea, but it went via the major town of Lincoln, and though some canal and much river management were needed, the Roman engineers and politicians must have thought it better than tackling The Fenland or making a canal to the east coast. It remained an important waterway link into the middle 1800s, after which much freight was transported by rail and then to the sea, the European continent, and on to the Mediterranean.

Lodes

Lodes Villages

Of unusual interest are the Lodes villages, all Roman (Fig. 42). Small (well, not as small or as lost as now) brooks ran off the chalk, but once down in flat land, the streams were dug and enlarged and became canals. The villages themselves were on dry land although house sites now by the Lode itself may presumably have flooded regularly in wetter times. Before modern utilities such as cables and pipes came into houses, furniture was (surprisingly) sparse, everything could be carried outside. Kitchenware and such-like were moveable. Flooding in the ground floor of houses was common. The same applied to houses in low valleys. The convenience of having water for domestic use, farm use, and transport often outweighed the inconvenience of occasional flooded houses.

A Lode village like Wicken could have the village itself on higher ground, with a walk down to the lode. It could have the village proper extending down to and on the side of the lode, as in Burwell which was noteworthy for its boat-building and trading. Warehouses and yards were still obvious there in the 1970s, though many were lost to new development by 2000.

Traditional wooden "cock-up" bridges enabled horse-drawn transport to cross the lode, the slats on the sloping ramps providing traction for the hooves. The bridge could be raised to let water traffic through. Such a bridge at Burwell, which could also be used for recreation, was mentioned in Eric Ennion's classic book, *Adventurer's Fen* (1942). Figure 43 shows a raised, modern cock-up bridge at Wicken Lode.

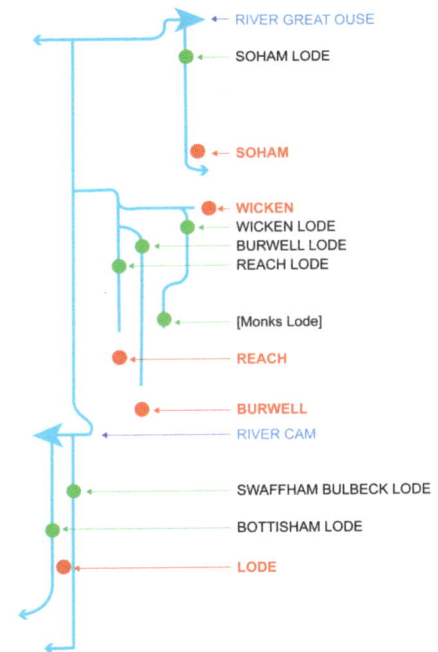

Fig. 42. Schematic diagram showing link-up routes for Cambridgeshire Lodes flowing to the River Cam and the River Great Ouse and the associated villages

Fig. 43. Modern cock-up bridge at Wicken Lode (April 2024)

Reach was famous for another reason—the annual Reach Fair held in May—where goods came from far and wide and were sold with much profit. The Fair dates back to 1201 when King John (1199–1216) signed its charter. Reach has a Green on the landward side which serves as a market place. Down on the flat land its layout allows ample space for a market place by the original lode too, and another for a channel presumably dug to accommodate increasing trading. Most of the old village pattern lies higher: the Fairs were held to fit local conditions (flooding and poor road surfaces in winter).

Lodes (Roman Canals)

Historically, along the southern edge of The Fenland, various tributaries flowed to the River Cam with river villages on the higher, drier ground, and canalised rivers along the fens known as Roman canals or Lodes (Fig. 44).

Today the picture is very different. Freshwater is running out. Few people understand this, but it is a very dangerous situation. All life requires water, land-life requires fresh water. The Law of Unexpected Consequences is written in stone.

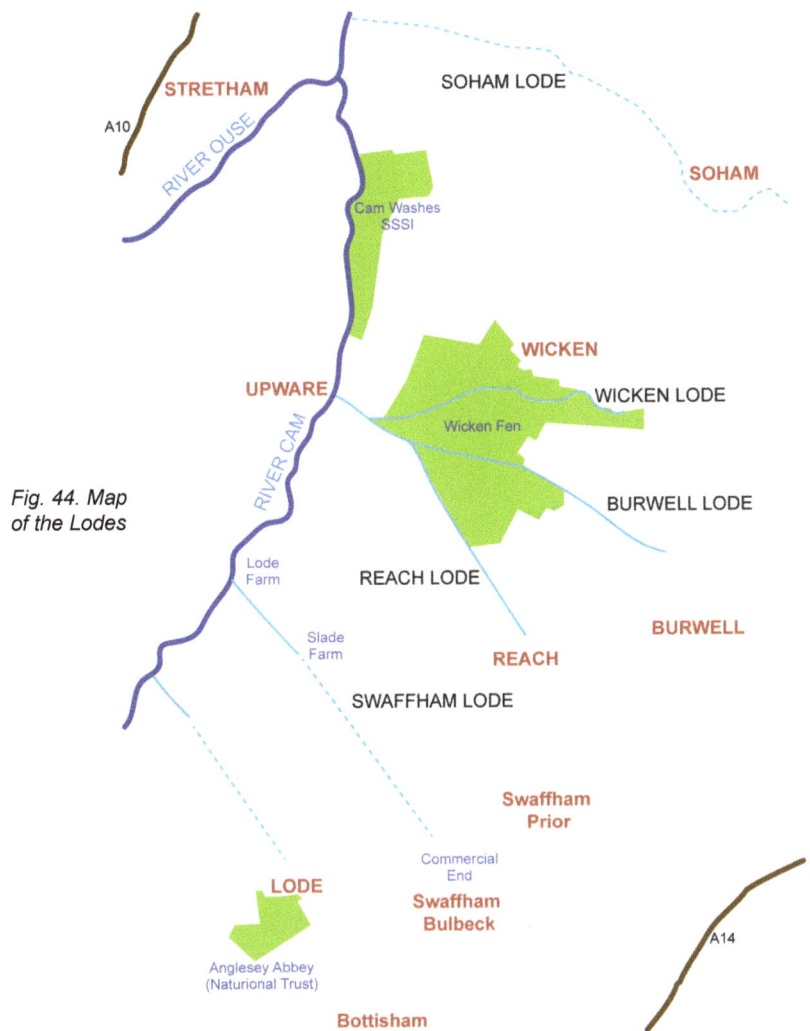

Fig. 44. Map of the Lodes

The efficient and privately owned Cambridge Water Company now have an ever-increasing water demand so also the streams have less water than before. (In 2025 the Cambridge Water Company operates as a "water-only" supplier under the ownership of South Staffordshire plc.)

The southern chalk ridge of Cambridgeshire (East Anglia *see* Fig. 26) is where most streams rise which supply the area lodes with water. The chalk ridge starts near the Rivers Lea and Mimram in the south around Hitchin and

63

Letchworth and runs centrally through Cambridge and on to Newmarket, Chippenham and Isleham and, laterally through villages in the four counties of Hertfordshire, Bedfordshire, Cambridgeshire and Suffolk.

Lode village itself is near Anglesey Abbey. It has a late eighteenth century mill, restored in 1868, which is still in operation. The whole mill complex has recently been further restored by the National Trust, and the area is now mainly used for recreation. It is not known when the near-by original pumping station was built, but the whole site was modernised in 2001 and its waterway was restricted to small craft such as canoes.

Reach Lode joins the River Cam at Upware. It was probably dug in Roman times and became an important mediaeval inland port. Visitors to its popular fair, established by King John's Charter in 1201, travelled by boat in to Reach to buy and sell their wares. The original Denver Sluice construction in the 1600s eventually prevented access by large ships but navigation by smaller vessels continued until the 1700s. Reach Fair still takes place every May Day Bank Holiday and has been held almost every year for over 800 years!

Swaffham Lode joins the River Cam at Lode Farm, running from just outside the village itself near Commercial End. This lode was used by small boats up to the late nineteenth century. Those in charge of the lode's upkeep installed several flood defences, including a lock gate, the height of which prevented the passage of larger boats from the Cam. In 2025 the lode is only navigable for about two miles from the River Cam upstream to Slade Farm.

Wicken Lode runs directly through Wicken Fen Nature Reserve, flowing to Commissioners Drain and on to the River Cam. This man-made watercourse is historically important being vital for the transportation of peat and sedge. The flow of water is now strictly controlled. The lode itself splits and runs through different areas of the fen. Wicken Lode is today mainly used for maintenance purposes, controlling the flow of water in and out of the reserve, and for recreation.

Burwell Lode: The village of Burwell was renowned for building boats, particularly the narrow boat, such as those used for The Grain Runs (*see* River Towns above) and later up to 1936 to transport general goods, including sugar beet right up to 1963. Burwell Lode is still navigable but is now used for recreation (boating and fishing), as shown in Figure 45, rather than for carrying goods. It is also very clear from the photograph that here the lode itself is higher than the surrounding fen countryside (*see* Fig. 5).

Fig. 45. Leisure Barge on Burwell Lode and fisherman on the bank (2023)

Incidentally, when Gordon Brown was Prime Minister (2000–2010) he decreed a new eco-town south of Cambridge. Alas, he did not understand that people in houses have a peculiar wish to have running water, and that the Cambridge Water Company was efficient. Result—no new eco-town. When Peterborough (on mainly clay, to the north west of the fens) decided to turn part of The Fenland back to fen, "The Great Fen", to demonstrate history and natural history, Cambridge, not to be outdone, announced the same. Alas, again, the Cambridge Water Company is still efficient and only a small, peat-walled area could be managed, not the large flooded fen envisaged for the project.

The "Great Fen Project" was finally conceived with five main supporting organisations comprising the Environment Agency, Huntingdonshire District Council, Middle Level Commissioners, Natural England and the Wildlife Trust for Bedfordshire, Cambridgeshire and Northamptonshire (BCNWT). In Autumn 2025 the BCNWT reported on their website and in their Newsletter that:

> *"The Great Fen is a landscape project to connect the two national nature reserves of Woodwalton Fen and Holme Fen by restoring the surrounding land for wildlife."*

> *"…The area is currently benefiting from one of the largest restoration projects of its type in Europe and the landscape is being restored and transformed for the benefit both of wildlife and of people."*

Vanishing Land

When peat dries it oxidises, so vanishes. Whilst present it stores carbon and anything else in the peat when it was laid down, such as the plants composing it, the pollen in the air, and any man-made objects that were buried or dropped—Excalibur, King Arthur's legendary sword, for instance, or pottery, spoons and other debris! **It is peat, being plant debris, which vanishes.** Inorganic material may dry and so shrink a bit.

In the north of The Fenland much more silt is deposited and this area is commonly known as silt fen rather than peat fen. Silt is inorganic and does not oxidise away like peat. So in the peat in the Cambridgeshire fens the land level has dropped substantially since the drainage of the fens in the mid-seventeenth century. Visitors, therefore, can be startled to find the River Cam tributaries flow well above the fields below. The Ship Inn, near Southery on the A10 is at the Great Ouse and Little Ouse confluence and clearly demonstrates the elevated waters and Inn surrounded by much lower farmland (*also see* Fig. 5).

The dykes and drains of the agricultural land now form a separate, controlled water system, with pumps here and there, such as the Haddenham Levels Pumping Station, lifting water to the river drainage system to flow to the sea (Fig. 46a,b,c).

Fig. 46a (May 2011). Haddenham Levels pumping station showing the dyke's entry point

Fig. 46b. Haddenham Levels pumping station discharging into the Old West River. The tempestuous water movement as the gates were released was joined by a thunderous roar, and the smell was rather stagnant too!

Fig. 46c. Discharge from the pump rippling the main river and forcing the foreground midstream vegetation to lean over in the water

Two pollution stories

1. Upstream of Cambridge at Harston

By the Second World War the mill at Harston had become an agrochemical factory owned by Fisons, which was later taken over by Bayer Crop Science. In the late 1940s, this company bought a lot of poison-gas precursor, hoping to turn it into useful fertilisers.

The Essex River Stour to the south was Drying Up (well, lessening water). More water was needed, as the River Stour tomato crop which grew there required a lot. What should be done? The idea of moving water via rivers was then new, but a new-fangled Water Transfer Channel was constructed *c.* 1970 to bring water from the River Cam to the River Stour and Essex reservoirs for irrigation. It worked well. BUT the tomatoes which were grown beside the Essex River Stour promptly died. Disaster! It turned out that the Harston factory, without knowing it, was emitting TBT. (Tributyltin—a toxic chemical used for various industrial purposes such as slime control in paper mills or disinfection of circulating industrial cooling waters.) This was only a few parts-per-billion—yes, parts per BILLION, not per million—but even beyond 20 miles downstream this was toxic enough to kill tomatoes. This was the first British indication that plants could be killed by such a low level of a pollutant, albeit, a highly dangerous pollutant.

In the early 1970s, research included looking at the River Cam downstream of the Harston factory. Indeed, plant diversity was low there, but it was assumed that this was due to disturbance by the pleasure boats (which, in part, it may have been). After a few years without TBT pollution, however, plant diversity went up. No one had thought previously to look at the plants AS WELL AS the invertebrates. To most people [wild] plants are just "dull weeds", whilst animals are far more interesting so the research focus was on these. But tomatoes are plants, so it raises the question that damage to tomatoes might also be relevant to all plants?

2. The River Cam in Central Cambridge

In the 1940s and 1950s, research into invertebrate pollution was carried out in central Cambridge. When the indices came in and were assessed, it was discovered that plants in the River Cam, with all the punts and other disturbance, were damaged and that the number of invertebrates was very low. This was recorded as "severe pollution". But was it? Unstable, disturbed silt may not be good for plant roots? What could be done? Many river invertebrates live on river plants. Real plants could not grow there, but what about fixing plastic plants in the river? Yes, that worked well. The invertebrate index improved sharply—because the river was so much better structurally, not because pollution and *real* aquatic plant growth was better. Because tests were developing, there was a sensible idea of improving habitat, so pollution effects could be separated from those of disturbance. Unfortunately the practice of "planting" plastic plants was later considered inappropriate so was stopped. It would have been much better, later, if artificial habitat had been standardised and structural- and chemical-damage factors had been measured separately, not always as combined factors.

Downstream of Cambridge

The River Cam is still Navigable to the sea via the River Great Ouse which it joins at Pope's Corner (*see* Fig. 2). Denver Sluice (*see* Fig. 3) was unfortunately not designed for sea-going boats, and transhipment to what became known as Fen Lighters developed. But having to unload, carry over, and re-load cargo to another ship made water transport far more difficult from the coastal ports. Traditional Fen Lighter boats were generally 30–40ft in length and 10ft wide. When empty, the draft was 2ft and they could carry 20–25 tons. Some were horse-drawn but later barges were strung together with a sturdy rope and were steam driven. They also had sails, and used water currents for propulsion. Fen Lighters were distinctive, oak-built river-crafts used between *c*. 1700 and the mid-nineteenth century, after which time transportation was largely replaced by the rail network.

Before 1996 there was still a "water-bus" service between Cambridge and Ely, that is, a time-tabled service by boat. The *Vicountess Bury* (built by W.S. Sargent & Co. of Chiswick, London in 1888) was the last boat to provide the service when she suddenly and mysteriously disappeared from the water and was never seen again! Water-buses are, and were, rare in England, though common in The Netherlands. In the eighteenth century the boats ran twice a week, taking 6 hours. In Ely there is a special, and grand, wharf for them where visitors can still enjoy a nice lunch inside The Cutter Inn pub or outside beside the river towpath. Boats are moored on the river Great Ouse which flows beside the quay.

In contrast, in Cambridge, there was not even access to the city centre, merely at the next lock downstream from the centre at Jesus Green. Here amid a plethora of pleasure boats the water-bus had to try to dock to allow passengers to disembark. "How are the mighty fallen!" This service used to be important!

The downstream section of the River Cam runs along and in the outer fenland which here is fen peat, laid down under wetlands which were alkaline and calcium-rich. Peat really started drying and disappearing with the seventeenth-century draining of the fens, when fields and crops replaced wetlands, fish and thatch. As draining technology improved, peat-loss accelerated in the nineteenth century and still more in the twentieth century. So the shrinking peat has meant the fens now start further into The Fenland than before, and the villages along the edge are less liable to flooding. Even between 1950 and 2000 the retreat of the peat was very noticeable. Generally, each farm had some fen land, used for peat for fuel, summer grazing, and crop-growing where the peat level was high enough.

In 1850 William Wells, a local landowner who foresaw land shrinkage, commissioned an oak "peat post" to be sunk at Holme Fen by engineer John Lawrence. In 1851 the oak post was replaced with a cast-iron column which was totally buried with just its pointed top at ground level. The "Holme Fen Post" became a Grade II listed monument in September 2020 and records show that about 13 feet (4m) of peat has been lost since 1851. The **bottom** of the post is now at ground level and the post itself is held up by several metal stanchions. Fenland vegetation has been replaced by silver birch with bracken undergrowth.

In the outer Fenland also are the line of Lodes (*see* Fig. 42) (Roman canals, rivers of "The Grain Run"), rising on the chalk, navigable from the villages. The many dykes and drains were all interconnected for boats, and connected up to the northern cut which, via such New Cuts and existing rivers as were available, allowed boats to travel up the Rivers Cam and Ouse, out into The Wash at Lynn, then in to the estuarine River Witham, on to the River Trent and hence to the sea at the Humber Estuary.

Before Ely is reached, a river of about equal size can be seen to the north of the River Cam, also flowing slowly and embanked. This is the River Great Ouse, or rather one branch of this much altered river, often known now as the Ely Ouse. These two rivers join at Pope's Corner (see Fig. 2), and from here the River Cam is subsumed into the River Great Ouse. Given the great importance of Cambridge, it could have been the other way: the coastal Cam river absorbing the water of the Great Ouse?

Ely

The Isle of Ely is the largest of the clay ex-islands in the southern, or peat fens. Administratively it is now incorporated into Cambridgeshire, as it gradually lost its importance (and more recently part of its [monastic] population) after the Dissolution of the Monasteries under Henry VIII in the sixteenth century.

The story of Ely is also the story of Etheldreda (Fig. 47), who died *c*. 673, unique amongst English women as being Princess (of Kent), Queen (of Northumbria), Abbess (of Ely) and Saint—her two sisters and niece were also canonised. Etheldreda was given the Isle by her first husband (the then "Prince of Fenmen"), settled there after her second marriage, and founded the Abbey. As was typical of the Anglo-Saxons, it was a double monastery (both monks and nuns and, as usual, headed by an Abbess). The Abbey was presumably destroyed by Vikings in around 870 and when re-founded around 970 it was just off-site and inhabited by Benedictine monks—no more ladies. What did the monks think? (Even now, no abbesses are named in the cathedral list!)

Fig. 47. St Etheldreda of Ely. Painting modelled on a full page miniature (Atheldrythe abbot and perpetual virgin), British Library MS Additional 49598 fol. 90v

Like Wells Cathedral, Ely is unique, though in a different way. Until the recent extreme drainage, mist rising from the wetland often used to obscure the lower part of the building, and the cathedral, apparently floating on cloud, affectionately became known as "The Ship of the Fens" (Fig. 48a).

Ely Cathedral prospered, became wealthy and the centre of a large and rich diocese, considered second only to Glastonbury. Its Octagon (Fig. 48b) is still a mediaeval wonder which could not now be repeated, as trees are too small for the building structure. This, and the castle, were built on the hill.

Fig. 48a. The ghostly outline of the "Ship of the Fens" shrouded in mist

Fig. 48b. Cathedral Octagon lit up October 2023 (Courtesy of Philip M. Ball MA, FMAA. University of Cambridge)

Ely was always limited by its being a remote island in a huge wetland and this of course affected the town's development. Today the market place, when not in use doubles as a car park. It is beside the Cathedral and has changed little over time. In the mid 1800s it was mainly used for the selling of livestock, and later was a clear space for recreation.

In 1897 a fountain was built to commemorate Queen Victoria's Diamond Jubilee (60 years), and the fencing round the market place was removed. *Circa* 1939, the commemorative fountain was taken from the market place and moved to Archery Crescent about 300m away. In the 2020s the market place has a thriving number of events. Ordinary markets are held every Thursday, Saturday and Sunday, with Farmers' Markets on the 2nd and 4th Saturday of the month, and Mini-Markets every Friday. As is usual in a church-dominated town, there is a direct way down the hill to the old commercial wharf and its associated buildings, including an old inn. Ely has recently redeveloped its trading past as pleasure boating, with a fine marina which boasts: *"Peaceful, fully serviced grassy moorings with extensive and varied cruising options. Ely Cathedral and the City Centre are on the doorstep"*. There are also good roads and a railway station.

Interestingly in the next county of Suffolk, the village of Icklingham on the Cam tributary River Lark, was on the dividing line between the diocese of Ely and Bury St Edmunds. Its Church of All Saints up on the hill recently became disused in favour of St James' in Icklingham village. It still conserves some of its old hassocks (kneelers), cut from the Great Tussock Sedge which still grows beside the river. It is now rare to find wetland products in river villages (Fig. 49)!

Fig. 49. Clockwise from left, Great Tussock Sedge (Carex paniculata); old sedge hassocks; sedge tussocks growing in wetland

Littleport

The second small town on the River Cam—the upstream one being Saffron Walden—is Littleport, and a "little" port it is, linked in to Ely. But it does have a well-constructed wharf with several moorings, as shown in Figure 50, situated alongside the very popular "Swan on the River" Restaurant and Hotel.

The River Great Ouse-Cam continues, much managed, on towards the sea, passing **Denver Sluice** (*see* Fig. 3). (For those brought up on Dorothy Sayers' *The Nine Tailors*, this structure and the eastern fenland are those described in the book, which is well worth studying for the evocative fenland, riverscape, and flood. But alas, there is no Duke of Denver or village of Duke's Denver.) With the land and sea levels as they were between 1500 and 2000, without some major removal of water, there was no expectation of a prosperous agriculture in the fens and the sluice has since been "re-built" several times.

Fig. 50. Wharf at Littleport by the "Swan on the River" Restaurant and Hotel, April 2023. Photograph courtesy of Paul Bone

In the seventeenth century, when there was much drainage and wealth in The Netherlands, a Company of Adventurers (investing—adventuring—their money and nothing to do with exploration) was formed to drain the fens, and a Dutchman, Cornelius Vermuyden, was employed. In The Netherlands it was already known that dry peat oxidised and vanished, lowering ground level. In 1651 the sluice and the dam on the Great Ouse were working, and there was much alteration to rivers, drains and dykes nearly all over the fens in Cambridgeshire and its surrounding counties. This prevented free passage of boats. Over time, repairs and refurbishment were needed, but agriculture in the fens is still very good, and farms are prosperous. However, as mentioned above, in places the peat has now gone and if it was over heavy clay, untreated crops grow less well.

Denver Sluice controls the Ely Ouse (incorporating the River Cam), the New Bedford River, the tidal flows and a connection to the River Nene (Peterborough and Northampton). Truly, in its way, also a "wonder" of the River Cam!

King's Lynn

A "linn" was an Anglo-Saxon estuarine pool. So it was good for extracting and exporting salt—the reason this small port first came to historic notice. The salt, and small port town of (then) South Lynn were well established by 1086, the time of the *Domesday Book*. The salt tippings for this industry raised ground level, which allowed the town to grow.

Whilst Cambridge was hindered from following standard development of the time by being "deformed" by Henry VI, King's Lynn (named South Lynn and Bishops' Lynn until the Dissolution of the Monasteries in 1537) took a different path. It was an unusual but not unique path also found, for example, in the German River Rhine towns in the Cologne region. Instead of developing on an original site, four or five different sites were developed in turn, adjoining downstream (Fig. 51a,b).

South Lynn Church—All Saints—with its Market Place is the first and oldest church and centre. It was limited by the rivers but the basic pattern can still be seen (Fig. 52). All Saints had an Anchoress for several mediaeval centuries and even in 2000 there was still a pervasive and peculiar feel to the church, from so much prayer for so long. (Anchoresses lived in a room (cell) at the side of a church to live a life of prayer and contemplation.)

Fig. 51a. Schematic Map of Lynn showing the market places and new streets that were springing up in the second quarter of the nineteenth century (c.1846). The railway station had just been built

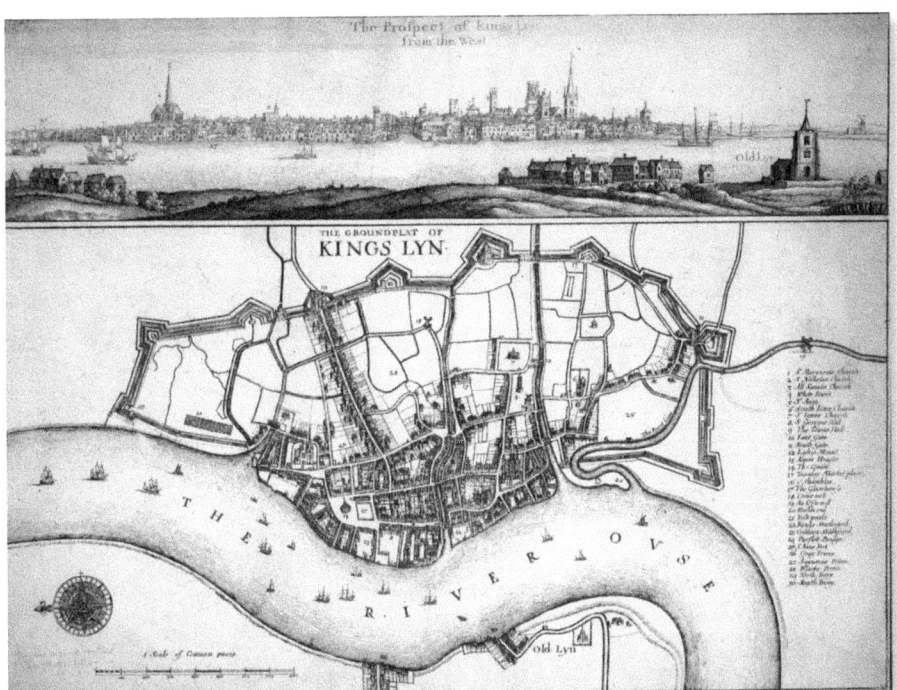

Fig. 51b. Top: part of Old Lyn's waterfront lined with warehouses into which small boats could gain access at high tide through watergates. Unusually high tides flooded cellars and houses!

Below: Groundplat of Kings Lyn. The green belt between the built-up area and the landward town defences gave Lynn the air of a garden city. Image courtesy of The Thomas Fisher Rare Book Library, University of Toronto

Next downstream is the St Margaret-and-Saturday Market Place town centre. This started in 1101 and was placed between the rivers Millfleet and Purfleet (compare with Fleet Street and River Fleet in London). This was grander and larger than the All Saints complex, and the "Important Buildings" such as the Custom House (the old building where you paid the accustomed dues; the plural, "Customs", came later), prison, Hanseatic League warehouses (Hanse House, St Margaret's Lane, King's Lynne, *c.* 1475), the friary, and so on, are all found in this area. The River Purfleet made an excellent harbour. In mediaeval times the Purfleet landing stage was much larger than it is today.

In the early thirteenth century this minor town became a major port, indeed the most important one in England in the fourteenth century. This change, like that in Cambridge, was due to an outside factor: floods had moved the main river entrance to the North Sea to here. (Before, it was further up the

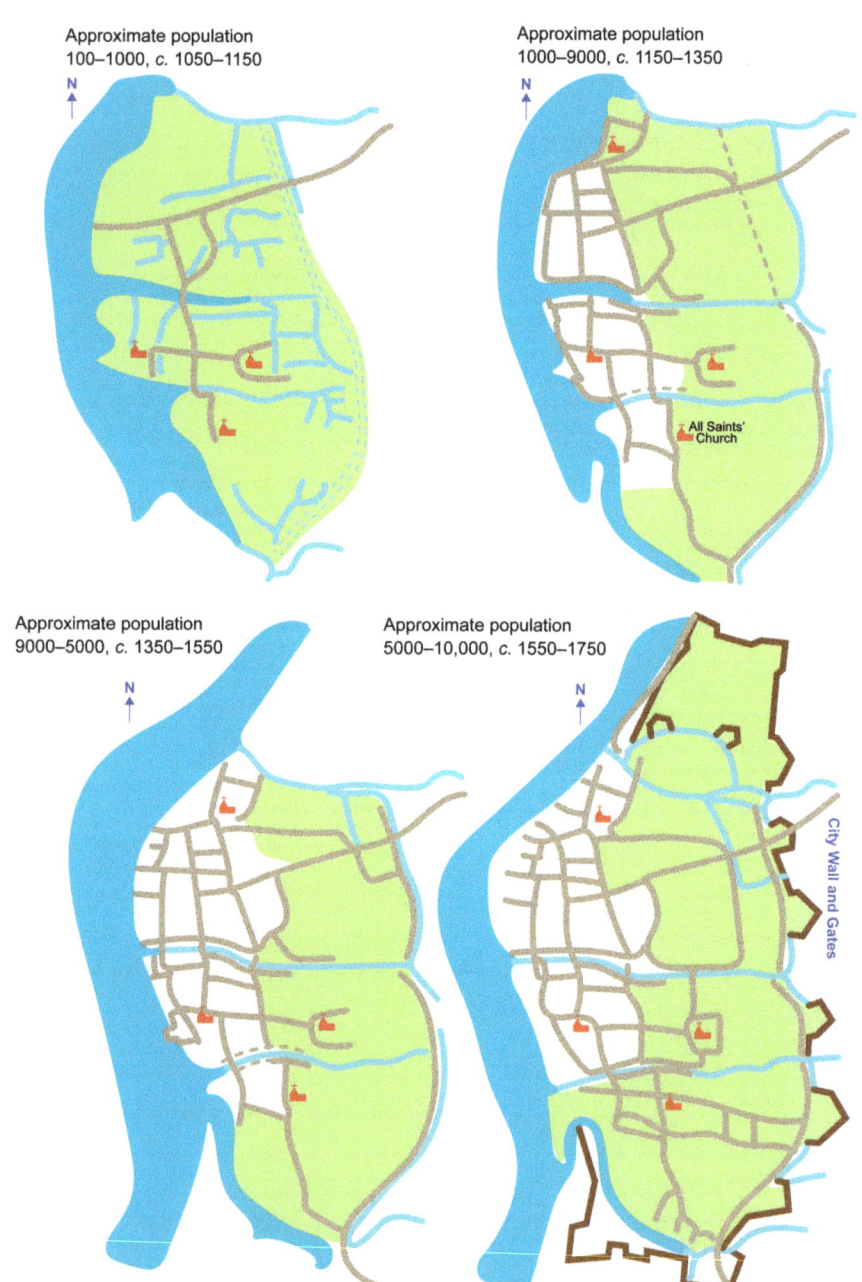

Fig. 52. Plans of early mediaeval Lynn showing development of roads and approximate population figures between 1050 and 1750. The latest official population figure, based on the 2021 census, is estimated to be 42,800 (source: Wikivoyage)

coast with the River Nene, taking in the central fens [Wisbech], Peterborough, Northampton and Daventry, and the River Nene waters.) So the gain was greater, and King's Lynn trading and prosperity greatly increased—the St Margaret's town centre shows this well! Lynn was now directly connected with Ely (centre of the diocese) as well as with Cambridge (County trading and after c. 1200, a University town) and the south Midlands. There was an important port in the now-small Millfleet which is still obvious at the site of the old inn. Lesser wharfs were on the less stable main Great Ouse. Shipbuilding and sea fishing became important industries. The housing, commercial and industrial development of King's Lynn as a whole increased greatly up to the eighteenth century.

With all this expansion, although there was development inland, instead of keeping the St Margaret's complex as the centre, a third Town Centre was sited, also alongside the main river because that is where the rich trading would occur in and around 1200 (quite soon after St Margaret's) and a larger, even grander "Tuesday" Market Place (now a car park) was built. Markets were a privilege and in theory were tightly controlled by either the King, Mayor and Corporation, or local magnates. Saturdays were already occupied by the St Margaret's market, so the new one was held on Tuesdays. Unlike the two earlier markets (All Saints, Millfleet and St Margaret's), the Tuesday Market Place did not have a new church of its own. Instead St Nicholas', off to the downstream side, became the local church—and St Margaret's, founded as the grander and main town church, remained so for ceremonial (and congregational) events. This St Margaret is the St Margaret of Antioch (Fig. 53).

Enough remains to show a typical, early to middle-mediaeval pattern. South Lynn was built in 1250 with a bridge connecting the St Margaret's complex, Bishops Lynn, with the older All Saints complex.

King's Lynn, however, necessarily remained remote, sometimes feeling more linked to the Dutch than to London. This meant it developed very much as a world within itself, with a wide range of enterprises, much wealth, and (local) importances.

A permanent check to the expansion of King's Lynn came with the discovery of the New World during the sixteenth century, and the consequent movement of part of Britain's trade to west-coast ports.

The development of the next "Town Centre"—hardly a Centre—was, not surprisingly, effectively delayed until the railways of the nineteenth century,

Fig. 53. St Margaret of Antioch, depicted on King's Lynn Town Sign (2015) with the motto: "in principio erat verbum"—"In the beginning was the Word"—the opening words of St John's Gospel

and did not receive grandeur. This fourth Centre may now be perceived as ugly and industrial, a sad come-down from the Second and Third Centres. No Church, no beauty: but intermittently much wealth.

The Millstream (with Mill!) is the westerly small stream, next after the less stable River Nar (*see* Fig. 2.) The area was described as "beautiful houses of habitation".

In Xanadu did Kublai Khan
A stately pleasure dome decree
Where Alph the sacred river ran
Through caverns measureless to man
Down to a sunless sea....

While none can rival Samuel Taylor Coleridge, is not the River Cam also a Riveting Riverscape?

Citations

Coleridge, S.T. *Kublai Khan (Xanadu)*, (1816). Full poem is available to view on the River Friend website:
http://riverfriend.tinasfineart.uk/resources/

Defoe, D. (1724–26). *A Tour Through the Whole Island of Great Britain* (English Library, 1962).

Ennion. Eric A. R. (1942). *Adventurer's Fen*. (ISBN: 9780905899411). Available on Amazon.

Haslam, S.M. (1991). *The Historic River – rivers and culture down the ages*. (Illustrations by Y. Bower). 324pp paperback, Cobden of Cambridge Press. (ISBN 0 9517963 0 5)

Haslam, S.M. (1997). *The River Scene – Ecology and cultural heritage*. CUP. (ISBN 0 521 57410 2 Hardback)

Lee, John Stephen, Corpus Christi College, Cambridge. "Cambridge and its Economic Region, 1450–1560". Dissertation submitted for the degree of Doctor of Philosophy 15th December 2000.

Old English Nursery Poem. Read whole version here:
https://riverfriend.tinasfineart.uk/resources/

Parker, R. *The Common Stream*. (1975). William Collins Sons & Co. Ltd., Glasgow. 283 pp. (ISBN 0 00 216113 3)

Sayers, Dorothy L. *The Nine Tailors* (1934). Gollancz, 350pp.

White, Rev. Gilbert (*c*. 1789). *The Natural History of Selborne*.

White, William. (1856). *White's History, Gazetteer and Directory of Lincolnshire, and the city and diocese of Lincoln*. White's Directories were a series of directory publications issued by William White of Sheffield, England, beginning in the 1820s.

List of Published Stand-alone Titles in the River Friend Series

THE RIVER FRIEND is a series of small books designed for people with a general or specific interest in rivers. Standalone* Titles in the Series include:

RFS1. *DRYING UP* (ISBN 978 1 9162096 1 9)

RFS2. *STREAM STORY I: A Riveting Riverscape—River Brue, Somerset* (ISBN 978 1 9162096 0 2)

RFS3. *A PROLOGUE TO THE SERIES: Plant identification and Glossary of Terms* (ISBN 978 1 9162096 2 6)

RFS4. *INTERPRET: What do Plants Tell us?* (ISBN 978 1 9162096 5 7)

RFS5. *REED—ON THE EDGE* (ISBN 978 1 9162096 4 0)

RFS6. *An Introduction to the WATER FRAMEWORK DIRECTIVE* (ISBN 978 1 9162096 3 3)

RFS7. *WATER: Clean and Dirty* (ISBN 978 1 9162096 7 1)

RFS8. *VEGETATION CHANGES OVER TIME. Is there freeze frame?* (ISBN 978 1 9162096 6 4)

RFS9. STREAM STORY II: Another Riveting Riverscape—River Cam, Cambridge (ISBN No. 978 1 9162096 9 5)

* Each book is about a different subject so the series can be read in any order

About the Authors

Sylvia Haslam is a specialist in river and wetland vegetation and river culture. Anyone wanting to find out more should look at the publications list on her website **(https://www.riversandreeds.co.uk)**. Her publications specific to this series are listed in RFS3: *A PROLOGUE TO THE SERIES: Plant identification and Glossary of Terms*.

Tina Bone has worked as a self-employed Desktop Publisher for many years until she changed career to work as a Professional Artist and Book Publisher from March 2005. To view Tina's résumé and artwork please visit her website: **https://www.tinasfineart.uk.**

www.ingramcontent.com/pod-product-compliance
Ingram Content Group UK Ltd.
Pitfield, Milton Keynes, MK11 3LW, UK
UKHW051014181225
466149UK00002B/5